T0271351

Climate Change and Eye Disease

This book examines the impact of climate change on eye disease and eye health.

Filling a lacuna in the existing literature, Scott Fraser takes a deep dive into the eye diseases that are most affected by the climate crisis and explores the subsequent burden on organisations, charities and healthcare systems. Fraser begins by including short primer chapters on the basics of climate science and climate change, highlighting which environmental mechanisms directly and indirectly affect our health and why. He then looks in detail at the direct and indirect threats to eye health from climate change and examines factors including changing insect vectors, trauma from extreme weather events such as wildfires, floods and droughts, as well as the impact of crop failure, malnutrition, animal and plant migration. Highlighting the Global North vs South divide, the book goes on to consider issues around eye care, exploring the increased burden that climate-induced chronic eye diseases including cataracts, macular degeneration and nutritional eye diseases are placing on health care systems. These chapters also reflect on the ways in which eye care, ophthalmology, optometry, pharmaceutical and medical device companies all contribute to the climate footprint themselves.

Unique and timely, this book will be a great resource for students and clinicians of ophthalmology, optometry and allied eye care professions, as well as climate scientists, researchers, policy makers, charities, NGOs working in related fields of environment and health.

Scott Fraser has been an NHS doctor for 35 years and an ophthalmologist for most of them. He has been active in research during this time including in the Cochrane Collaboration. He was Visiting Professor at the University of Sunderland in the UK and the University of New England in Australia. As well as research papers he has written a number of books most recently 'Your Health When the Climate Changes.'

Routledge Studies in Environment and Health

The study of the impact of environmental change on human health has rapidly gained momentum in recent years, and an increasing number of scholars are now turning their attention to this issue. Reflecting the development of this emerging body of work, the *Routledge Studies in Environment and Health* series is dedicated to supporting this growing area with cutting edge interdisciplinary research targeted at a global audience. The books in this series cover key issues such as climate change, urbanisation, waste management, water quality, environmental degradation and pollution, and examine the ways in which these factors impact human health from a social, economic and political perspective.

Comprising edited collections, co-authored volumes and single author monographs, this innovative series provides an invaluable resource for advanced undergraduate and postgraduate students, scholars, policy makers and practitioners with an interest in this new and important field of study.

The Politics of the Climate Change-Health Nexus
Maximilian Jungmann

Research Ethics for Environmental Health
Edited by Friedo Zölzer and Gaston Meskens

More-than-One Health
Humans, Animals, and the Environment Post-COVID
Edited by Irus Braverman

Covid-19 and Global Inequalities
Vulnerable Humans
Victor Jeleniewski Seidler

Climate Change and Eye Disease
Eye Health and Blindness in a Hostile Environment
Scott Fraser

For more information about this series, please visit: https://www.routledge.com/Routledge-Studies-in-Environment-and-Health/book-series/RSEH

Climate Change and Eye Disease

Eye Health and Blindness in a Hostile Environment

Scott Fraser

First published 2025
by Routledge
4 Park Square, Milton Park, Abingdon, Oxon OX14 4RN

and by Routledge
605 Third Avenue, New York, NY 10158

Routledge is an imprint of the Taylor & Francis Group, an informa business

British Library Cataloguing-in-Publication Data
A catalogue record for this book is available from the British Library

ISBN: 978-1-032-84408-4 (hbk)
ISBN: 978-1-032-84409-1 (pbk)
ISBN: 978-1-003-51260-8 (ebk)

DOI: 10.4324/9781003512608

Typeset in Times New Roman
by Deanta Global Publishing Services, Chennai, India

Contents

Introduction

Our climate is changing and human activity is the cause. There is no reasonable doubt for this and the evidence continues to stack up. If you are unsure of this evidence then it is briefly discussed in Chapter 1, but there is no shortage of sources of evidence or studies available. If you are unsure what these changes in our climate are doing to us humans, then this is outlined in Chapter 2. We can partially adapt to our warmer world, but when circumstances become so extreme and we are no longer able to do this, our health is threatened. As we will discuss, more dangerous and insidious are the less direct effects our changing climate has on us. From crop failures to antibiotic resistance to conflict over food and water, climate change slowly creates threats in all our daily lives.

This book, however, is about a particular threat to human health – a threat to that of the eyes and visual system. 70% of the sensory input to our brains comes from our eyes. This is because good vision is essential for us to survive in the world – from obtaining food to escaping predators. In our modern world, blindness and visual impairment are now less likely to be life-threatening but carry a substantial burden for the individual and the society they live in. Our eyes are directly exposed to the external environment and so are directly exposed to the increasing hazards of climate change. We will discuss this in detail in Chapter 3.

How will climate change affect our eyes and vision? This is discussed in Chapter 4. The mechanisms are generally the same as those threatening our overall health such as increased heat, extreme weather, air pollution, droughts and floods, and changes to infectious disease patterns and vectors. One risk factor that our eyes share with our skin is that of ultraviolet radiation damage. As atmospheric chloroflourocarbons (CFCs) decline some of the protective ozone layer is recovering but evidence indicates that the burning of fossil fuels can reverse this recovery.

Chapter 5 looks at the damage that eye care in all its forms does to the climate. From the greenhouse gas emissions of healthcare faculties to transport, from single-use plastics to poor disposal systems, the climate footprint is heavy. This has begun to be addressed and eyecare organisations and practitioners are thinking more and more about the harm they do to the world.

DOI: 10.4324/9781003512608-1

Crucially, tools to allow comparisons between units have been developed and allow us to see how much more waste occurs in US and European eye facilities compared to those in Asia and Africa.

Chapter 6 looks at the burden of blindness and visual impairment worldwide. We know this is already substantial, but if the climate predictions outlined in the previous chapters are correct then this will become much worse. The prospect of more and more sight impaired people in an increasingly hostile world is frightening.

Finally in Chapter 7 we look at some hope for the future. As mentioned above organisations and individuals are beginning to think about the climate and what can be done to protect it. Sustainability is increasingly becoming as important as efficacy and costs – as it has to. We also need to address the eye diseases that currently plague our world in readiness for the escalation of threats we will face. In some parts of the world and for some diseases (e.g. trachoma, Vitamin A deficiency) successful campaigns have reduced the number of cases and therefore preserved many people's sight. The lessons we have learnt from this could be very useful in the future.

Whilst I have tried to be as comprehensive as possible, it is not possible to cover every effect of climate change on our eyes. Similarly, I have tried to use the most illustrative statistics rather than offer long lists of numbers which can ultimately hide a message rather than illuminate it. I have addressed the main eye diseases for which we already have evidence of environmentally induced damage and those which are most vulnerable in our warmer future. Conversely, I have addressed the effects of eyecare on climate change using the evidence we already have and as time goes on more evidence will accumulate. More research is beginning to be funded and it is very important this accelerates.

Before I started this book I had little doubt that climate change is affecting every part of nature – including the large part that we humans occupy. The scale of the threat to our health and, by extension, to our vision is even greater than I thought. We all need to act and, as healthcare practitioners, we have the knowledge and the platform to lead. We must do so before it is too late.

1 Why is our climate changing and what is the cause?

1.1 What is the evidence that the Earth is getting hotter?

It is hard to find better evidence for climate change than that contained in The *Intergovernmental Panel on Climate Change* (IPCC) reports. These were first issued in 1988 and most recently in 2023 [1].

The 2023 IPCC report is clear that between the years 2011–2020 the average global temperature is 1.1°Centigrade (°C) higher than between 1850 and 1900. Of this increase, 1.59°C is land temperature rise and 0.88°C sea temperature rise. Global surface temperature has risen faster since 1970 than in any 50 year period over the last 2000 years [1].

The joint Royal Society (UK) and the US National Academy of Science *Climate Change Evidence and Causes* [2] in 2020 provides an equally worrying picture. Half of the surface temperature increase has been since the mid-1970s. The lower and upper layers of the oceans have all warmed. Snow and ice are decreasing significantly and sea levels rising. It is estimated that since 1902 sea levels have risen around 16cm and between 2010 and 2020 the average global rise in sea level was 3.6mm. 2023 was the warmest year ever recorded [3].

These are numbers spread over time and they can sometimes be difficult for us to connect to people's daily lives. But what is obvious is that all over the world temperature and weather records are being broken and that living with higher temperatures is increasingly becoming the norm. This is outlined in the *2022 Report of the Lancet Countdown on Health and Climate Change* [4]. During 2021–2022, record temperatures were recorded in Australia, Canada, India, Italy, Oman, Turkey, Pakistan, and the UK. Temperatures exceeded 50°C on 16 July in Death Valley in the US. China's national temperature record was shattered on 16 July 2023 at 52.2°C (at this temperature humans could not survive outdoors for long). The following day, the US had its highest ever nighttime temperature in Arizona, which also set its record for the longest time without the temperature falling below 90°F/32.2°C [5].

Also in July 2023, temperatures of over 40°C were measured in parts of Spain, France, Italy, and Greece. Sicily saw temperatures as high as 115 degrees (46.3°C). In Rome, a new record high temperature of 41.8°C was recorded on 18 July.

DOI: 10.4324/9781003512608-2

As well as being directly hazardous to humans, these extreme temperatures can cause other extreme events. Wildfires occurred more frequently in 61% of countries in 2018–21 compared to 2001–4. This has resulted in a yearly average wildfire exposure increase of over 9 million person-days [4]. Droughts have become more common: in 2012–21 47% of global land area was affected by extreme drought annually, but between 1951–60 this number was only 29%. The Middle East and North Africa were particularly affected, with some areas having an extra 10 months of extreme drought. [4]

The list of statistics is almost endless, and they all point in the same direction. The effects of this extra heat (the extra energy) on our planet are profound and disturbing – from hurricanes and tornadoes to earthquakes, floods, ocean acidification, changes in plants, insects and micro-organisms and changes in large animal behaviour.

The delicate balances of all the Earth's ecosystems are being profoundly affected and this, as we will see in later chapters, is beginning to significantly affect human health. We have evolved to live on an Earth which is now changing too rapidly for our bodies to adapt to.

1.2 How do we know the Earth is getting hotter?

Historical evidence unequivocally shows rising global temperatures. Direct measurement of atmospheric temperature began in the 1850s and continues to the present day, in which we have multiple recordings the world over. These measurements provide direct evidence of the increasing temperatures over the last 175 years [6].

Using tree rings and ice core samples we can go even further back in time and get an idea of temperature changes over millennia [7]. It has been calculated that 1989–2019 was the warmest 30 year period in more than 800 years. 3–5 million years ago the average global temperature was about 2 to 3.5°C higher than in the pre-industrial period. At 50 million years the average temperature of the Earth was 10°C warmer than it is now. The Earth had little or no ice, sea levels were 60 metres higher than they are now. [2]. Thus we are heading back in time to a much warmer world in which humans did not exist.

1.3 What is causing the Earth to get hotter?

The science underpinning this rise in temperature is well-established – it is the increase in so-called greenhouse gases in our atmosphere and in particular carbon dioxide (CO_2). The detailed physics is beyond the scope of this book but it is vitally important to understand the concept. Gases in our atmosphere – especially CO_2 and water vapour – are essential for our existence on a planet that would otherwise be too cold. They allow the heat from the Sun to pass through but when this heat is reflected back by the Earth they prevent its escape into space. Thus they act like a 'greenhouse' and allow the Earth's life systems to flourish.

Measurements show that the atmospheric CO_2 level in 2019 was 40% higher than it was in the 19th century [2]. Most of the increase has been since 1970. The IPCC report is clear – in 2019, atmospheric CO_2 concentrations (410 parts per million) were higher than at any time in at least 2 million years and concentrations of other greenhouse gases (GHGs) such as methane (1866 parts per billion) and nitrous oxide (332 parts per billion) were higher than at any time in at least the last 800,000 years [1]. Humans are living on a planet with an atmosphere we have never previously encountered. We have no way of truly knowing the effect this will have on us or the organisms we share the planet with.

Where are these extra GHGs coming from? As *Climate Change Evidence and Causes* reports, the observed pattern of tropospheric warming and atmospheric cooling that measurements show could only occur with CO_2 from human activities [2]. If our rising global temperatures were related to increased output from our Sun, both the troposphere *and* the stratosphere would be equally warmed. Computer modelling of atmospheric data fits with CO_2 emissions from the burning of fossil fuels, rather than solar activates, volcanic emissions, or natural weather variations such as *El Nino* and *La Nina,* causing global warming.

If more evidence were needed, atmospheric CO_2 has been shown to have an 'isotopic fingerprint'. The type of carbon isotopes dominant and increasing in our atmosphere can only come from ancient terrestrial plant matter (i.e. coal, oil, and gas). Fossil fuels are the only source of carbon consistent with this isotopic fingerprint of the carbon present in today's atmosphere. [8].

1.4 What are the consequences of these changes in our atmosphere?

Higher temperatures and higher levels of CO_2 in our atmosphere have profound effects upon all the organisms inhabiting our planet. Extra heat in the atmosphere increases the likelihood, strength, and duration of extreme weather events. The scale of hurricanes, tornadoes, and typhoons are increasing. Flooding events are becoming more common and more devastating. Similarly, droughts are becoming more frequent, more widespread, and more prolonged. The *United Nations Office for Disaster Risk Reduction* (UNDRR) report indicated that, as of 2020, extreme weather events have come to dominate the disaster landscape in the 21st century. In the period 1980–99 there were 3,656 extreme weather events, but between 2000 and 2019, this increased to 6,681 events. Over the same periods, major flood events doubled, from 1,389 to 3,254, while the incidence of storms grew from 1,457 to 2,034.

The report also confirmed major increases in other extreme weather categories – droughts, wildfires and extreme temperature events [9].

Heatwaves are becoming more common and are lasting longer. For example, in the US, heatwave frequency has increased steadily from an average of two heat waves per year during the 1960s to six per year during the 2010s and 2020s [10]. The average heatwave season has increased by around 49 days [10]. Heatwaves are becoming more deadly and this is projected to increase [11]. They are also becoming more common in more northerly areas [12, 13].

As a result of these temperatures, wildfires are becoming more common, lasting longer, and becoming more deadly. These have doubled in number in the Western US between 1984 and 2015 and over the last decade there has been an unusually large number of wildfires even in Alaska [14]. Forest fires now result in 3 million more hectares of tree cover loss per year in 2023 compared to 2001 – an area roughly the size of Belgium – and accounted for more than one-quarter of all tree cover loss over the past 20 years [15]. Wildfires do of course have an added sting in their tail, as they pump huge amounts of greenhouse gases into the atmosphere.

Our oceans are being directly affected [1]. Global mean sea level increased by 200cm between 1901 and 2018. The average rate of sea level rise was 1.3mm per year between 1901 and 1971, increasing to 1.9mm per year between 1971 and 2006 and further increasing to 3.7mm per year between 2006 and 2018. The oceans are also taking up some of the extra atmospheric CO_2. This is one buffer against climate change but unfortunately the ability of the oceans to do this has been overcome. As dissolved CO_2 is acidic, the pH of our oceans has reduced. This has profound effects upon the inhabitants of those oceans – from damaged shells of shellfish all the way up the food chain.

Loss of ice from the polar regions and from glaciers directly affects the circulation of our seas. Nutrient cycling is altered in these oceans, which has profound effects upon the creatures living in the seas [16]. Major river systems of the world are fed by glacier melt from the mountains. The shrinking of these glaciers will directly affect these rivers and mean drought for millions downstream. A major example of this is the Himalayan glaciers which feed the great rivers of Asia (the Ganges, Indus, Brahmaputra, Mekong, Thanlwin, Yangtze, and Yellow Rivers). It has been calculated that these glaciers have lost around 40 per cent of their area – shrinking from a peak of 28,000km^2 to around 19,600km^2 in 2021 [17]. If this rate continues the many millions of people who rely on these rivers will be severely affected.

As plants rely on CO_2 to photosynthesise, it might be expected that they would thrive with raised levels in the atmosphere. To some extent this is true, and raised levels of CO_2 and warmer temperatures can increase the rate of photosynthesis. However, other factors are needed for this, such as nitrogen from soil bacteria, and this is disrupted by climate change. Climate change also means growing seasons are longer and warmer, allowing for greater plant activity, but this means more water is taken from the soil – increasing the risk of droughts. Longer, warmer seasons can also mean greater opportunities for pests to attack plants. Higher temperatures and other stresses can reduce plant yield and change their nutritional status and this can have significant impacts on human health.

Similarly, climate stresses can change animal behaviour. Pollinating insects can be lost or disease carrying insects can migrate to new areas and attack plants and livestock previously unexposed. Large animals migrate to cooler climes and thus can be lost as a foodstuff (e.g. fish) or can affect animals already present (e.g. predators of livestock). They may also compete for space and water with humans who were already present or who themselves have migrated. The human migration journey has its own risks and fatalities and the number of fatalities is increasing [18].

The list of adverse effects of climate change is very long and it is hard to find an area it will not impact. As we will see in subsequent chapters, humans are directly and indirectly affected by these changes. Direct effects from the atmospheric energy such as wind events create obvious health risks to humans. Less obvious is the changes in nutrition from plant and animal stresses and from disease vectors migrating to cooler climes. The full ramifications of climate change we do not, as yet, fully know, but we do already know that not a single area of our lives will be untouched by it.

References

1. The Intergovernmental Panel on Climate Change. AR6 Synthesis Report Climate Change 2023. https://www.ipcc.ch/report/ar6/syr/.
2. The Royal Society and the US National Academy of Sciences. Climate Change Evidence and Causes 2020. https://royalsociety.org/-/media/education/teacher-consultant-resources/climate-change-evidence-causes.pdf.
3. National Oceanic and Atmospheric Administration 2024. 2023 was the World Warmest Year on Record. https://www.noaa.gov/news/2023-was-worlds-warmest-year-on-record-by-far.
4. Romanello M, Di Napoli C, Drummond P et al. The 2022 Report of the Lancet Countdown on Health and Climate Change: Health at the Mercy of Fossil Fuels. *The Lancet* 2022:400;1619–1654. https://doi.org/10.1016/S0140-6736(22)01540-9.
5. World Weather Attribution July 2023. Extreme Heat in North America, Europe and China in July 2023 Made Much More Likely by Climate Change. https://www.worldweatherattribution.org/extreme-heat-in-north-america-europe-and-china-in-july-2023-made-much-more-likely-by-climate-change/.
6. North American Space Agency. Evidence. https://science.nasa.gov/climate-change/evidence/.
7. Climate.gov 2022. Past Climate. https://www.climate.gov/maps-data/climate-data-primer/past-climate.
8. Climate.gov 2022. How Do We Know the Build Up of Carbon dioxide in the Atmosphere is Caused by Humans? https://www.climate.gov/news-features/climate-qa/how-do-we-know-build-carbon-dioxide-atmosphere-caused-humans.

9. United Nations Office for Risk Reduction 2020. The Human Costs of Disasters. https://www.undrr.org/publication/human-cost-disasters-overview-last-20-years-2000-2019.
10. United States Environmental Protection Agency 2022. Climate Change Indicators: Heat Waves. https://www.epa.gov/climate-indicators/climate-change-indicators-heat-waves#:~:text=Heat%20waves%20have%20become%20more,threshold%20(see%20Figure%201).
11. Lüthi S, Fairless C, Fischer EM et al. Rapid increase in the risk of heat-related mortality. *Nature Communication* 2023;14:4894. https://doi.org/10.1038/s41467-023-40599-x.
12. Met Office 2020. UK and Global Extreme Events: Heatwaves. https://www.metoffice.gov.uk/research/climate/understanding-climate/uk-and-global-extreme-events-heatwaves#:~:text=The%20AR6%20report%20states%20that,intensity%20and%20duration%20of%20heatwaves.
13. Powis CM, Byren D, Zobel Z et al. Observational and model evidence together support wide-spread exposure to noncompensable heat under continued global warming. *Science Advances* 2023;9(36). https://doi.org/10.1126/sciadv.adg92.
14. Wuebbles DJ, Fahey DW, Hibbard KA, Dokken DJ, Stewart BC, Maycock TK (eds.). U.S. Global Change Research Program, Washington, DC, USA, 470 pp. https://doi.org/10.7930/J0J964J6.
15. Tyukavina A, Potapov P, Hansen MC et al. Global trends of forest loss due to fire from 2001 to 2019. *Frontiers in Remote Sensing* 2022;3. https://doi.org/10.3389/frsen.2022.825190.
16. Tuerena RE, Mahaffey C, Henley SF et al. Nutrient pathways and their susceptibility to past and future change in the Eurasian Arctic Ocean. *Ambio* 2022 Feb;51(2):355–369. https://doi.org/10.1007/s13280-021-01673-0. Epub 2021 Dec 16. PMID: 34914030; PMCID: PMC8692559.
17. Lee E, Carrivick JL, Quincey DJ et al. Accelerated mass loss of Himalayan glaciers since the Little Ice Age. *Scientific Reports* 2021;11:24284. https://doi.org/10.1038/s41598-021-03805-8.
18. Sindall R, Mecrow T, Queiroga AC et al. Drowning risk and climate change: A state-of-the-art review. *Injury Prevention* 2022;28:185–191. https://injuryprevention.bmj.com/content/28/2/185#ref-61.

2 How does climate change affect human health?

As we saw in the previous chapter, climate change will affect every corner of our planet, every ecosystem, and eventually every organism. Like our bodies, these ecosystems are finely balanced and have evolved to thrive within very limited parameters. Increasingly, our bodies and the systems we live in will be stressed by the changes in our climate and inevitably this will have an impact upon human health.

We can think of these impacts and threats to human health in two broad categories – direct effects and indirect effects. Direct effects are obvious; the indirect effects are less so, but likely to be cumulatively more deadly. Ten people being killed by a tornado makes the nightly news but 10,000 people slowly dying of malnutrition from reduced vitamins and minerals in stressed plants does not.

2.1 The direct effects to human health from climate change

Wind events

As greenhouses gases accumulate in our atmosphere, more and more of the Sun's energy stays within the atmosphere. Our weather systems are driven by this energy, and so as more energy accumulates, we witness more and more extreme manifestations. Perhaps the most obvious of these are the wind events – hurricanes (called typhoons in East Asia) and tornadoes. An atmosphere with more heat and greater amounts of evaporated water with warmer land (tornadoes) or warmer seas (hurricanes) allows these great forces of nature to generate. Though there has been some debate as to whether these wind events are becoming more numerous and longer lasting, there does seem to be evidence that they are more destructive [1] and that this destruction is likely to worsen [2].

The threat to humans is obvious either through collapsed buildings or flying debris. Less obvious but equally damaging threats come from the aftermath – from exposed electrical wires to damage to health facilities to lack of clean water supply.

DOI: 10.4324/9781003512608-3

Heat events

High ambient heat and heatwaves can cause significant mortality and morbidity. The cells and organs in the human body work within a very narrow range of temperatures and quickly degenerate when outside this range. The body's external temperature varies enormously but the inner temperature of the body – its core temperature – does not.

The hypothalamus is the part of the brain that is responsible for maintaining a very strict core temperature. When the external temperature begins to rise the receptors in the skin detect this and feedback to the hypothalamus. The hypothalamus is able to reduce the metabolic activity in the body and this reduces internal heat production. It also tells the skin's sweat glands to produce more sweat which takes heat from the skin surface as it evaporates. Similarly, increased blood flow to the skin allows greater heat loss via the air. The hypothalamus also tells our conscious brain to search for a cooler environment – going inside, finding shade, or jumping into a river.

However, when temperatures become too hot or last for too long this can be too much for the body to compensate for. In these circumstances the core temperature begins to rise and damage to cells and organs begins. If this damage is mild and normal temperature is quickly re-established, full recovery will occur. If the source of heat continues then severe heat-related illness (commonly called heat stroke) can develop and this can be fatal within a few hours. Even if death does not occur, survivors are often left with kidney, liver, brain, or other permanent organ damage.

As we saw in the previous chapter, heatwaves are becoming increasingly common.

Wildfires

One particularly nasty effect of prolonged high temperatures are wildfires. Although spontaneous fires do occur in nature (usually caused by lightning) these cover small areas. With higher and higher temperatures and dried vegetation, huge wildfires are becoming more common and occurring further and further north.

There is an immediate and obvious danger from wildfires and all living creatures in their path. However, the danger is not limited to those just in the immediate vicinity and the air pollution from wildfires contains a huge number of potentially toxic substances.

Wildfire smoke can travel thousands of miles, potentially resulting in widespread adverse health impacts. The smoke from a 10,000-hectare wildfire can affect people living in an area 10 to 15 times larger, impacting people who have never even seen the flames [3]. Most of the damage to health comes from breathing in the smoke with resulting respiratory problems. For those already with lung diseases such as asthma or chronic obstructive airways disease (COPD) this smoke is particularly dangerous. A longitudinal study conducted in Brazil showed that a 10μg/m^3 increase in wildfire-related particulate matter was associated with a 1.7% increase in *all-cause* hospital admissions, a 5% increase in *respiratory* hospital admissions and a 1.1% increase in *cardiovascular* hospital admissions between 0 to 1 day after the exposure [4]. Estimates from studies in 749 cities across 43 countries suggest that wildfire smoke and its associated air pollution directly cause over 33,000 deaths annually with an increased risk of all-cause mortality (i.e. it is not just the lungs it damages) [5].

Flood and drought

A warmer planet means more water evaporation from its surface. The warmer atmosphere can hold larger volumes of water vapour but eventually this has to fall. If it falls over a short time in a significant volume and onto land that has been dried out with heat then this water will not be able to drain quickly enough and flooding occurs. This is particularly so in urban areas where rivers might have been squeezed in town and city centres and where green areas have been replaced with car parking spaces.

Flooding again presents an obvious risk from the acute event – drowning, landslides, or trauma from floating objects. It also presents more indirect threats to health which we will deal with in the next section.

Droughts play out over a longer time period but kill far more people than floods do. The causative mechanisms are the same, though, with increased surface temperatures causing increased evaporation and water loss from rivers, lakes, and the soil. This water will eventually go somewhere, but not necessarily where it is needed.

Droughts most obviously limit the water supply to all organisms. Plants die, so the animals that graze on them die and humans starve. Droughts destroy the soil structure allowing landslides and prevent the soil absorbing water when it does rain. Static water in dried rivers or lakes encourages insect vectors that carry malaria to thrive. Poor water supply, and therefore poor hygiene, can result in a host of infectious diseases that we will discuss in the next section and in later chapters.

Less immediately visible than reduced rainfall is loss of ice and snow in the mountains. This doesn't just mean more expensive alpine holidays but represents a huge threat to millions of people. In many areas of the world, summer melting of ice and snow on mountains feeds the rivers. If these rivers dry up, the land of many, many millions of people would be in severe drought.

Those who live in coastal regions of the world (around 10% of humans) are at particular threat from climate change. Higher seas – from polar ice loss and from thermal expansion of the water – are an obvious threat. Hurricanes of greater power will mainly affect coastal communities. Storm surges (the rise in the sea level purely caused by a storm) are a particular hazard as they can be very high and powerful.

The resulting coastal flooding can salinise the soil, making it unusable for years. Changes in ocean acidity decimate shellfish populations. Warming seas cause fish species to migrate, leaving fishermen to travel greater distances in rougher seas to maintain their livelihood.

2.2 The indirect effects to human health from climate change

The threats to human health from climate change come in many forms. Such is the all-pervading nature of this change that it isn't possible to identify an aspect of our lives that will not be affected. Some of the major examples are outlined below but there are other threats to human health from our warming planet. Many we know about, some we can predict but don't know how severe they will be and other threats will unexpectedly emerge.

Perhaps the most worrying of those, and one already with us, is food insecurity. Heatwaves, floods, and droughts all affect plant and livestock health and survival. Famine is present in our world already and though often caused by politics and war, it will increasingly be caused by climate change (and climate change itself will be an increasing cause of conflict and war as populations migrate and natural resources shrink).

Even without the devastating effects of famine, the nutrition we gain from plants can be reduced when they are stressed. Although some crop yields may increase with rising atmospheric CO_2, unfortunately rising CO_2 levels can also reduce the level of important nutrients in crops [6]. With elevated CO_2, protein concentrations in grains of wheat, rice and barley and potato tubers decrease by 10 to 15% [7]. Crops also lose important minerals, including calcium, magnesium, phosphorus, iron, and zinc. A 2018 study of rice varieties found that while elevated CO_2 concentrations increased vitamin E, they resulted in decreases in vitamins B1, B2, B5 and B9 [8].

CO_2 levels expected in the second half of the 21st century will reduce the levels of zinc, iron, and protein in wheat, rice, peas, and soybeans. Some two billion people live in countries where they receive more than 60 percent of their zinc or iron from these types of crops. Deficiencies of these nutrients has already caused an estimated loss of 63 million life-years. [9].

As we saw above, coastal communities who gain most of their protein from fish and shellfish will be at risk as these sources die or migrate.

A warming world will alter the behaviour of insects and thus the diseases these vectors carry. Many insect species are vital in pollinating plants and if these are lost then food supplies are threatened. Disease carrying insects may flourish in a warmer world – many rely on static water to reproduce. As with other insects, those who carry disease may migrate to more suitable conditions and already we are seeing previously tropical diseases moving north [10].

Climate change creates a host of opportunities for infectious diseases to flourish. Disruption to water supplies causes poor hygiene, allowing diseases such as cholera and typhoid to flourish. Communities forced together in smaller areas increase the risk of passing on a range of infections. Reduced host immunity secondary to poor nutrition and health allow even relatively harmless infections to be life-threatening.

Micro-organisms themselves can be affected by climate change. Bacteria and fungi will change their behaviour, location, and even their biology under the pressure of climate change. Because of their relatively simple structures and rapid reproduction they can adapt to new environments much more quickly than larger organisms. These adaptations will have profound effects upon human health as we are presented with new diseases and a lessened ability to treat them. Although viruses are not directly affected by global warming they will have greater opportunities to infect us. Once again, disruption to water and land and population pressures will increase the risk of infections. Poor nutrition lowers resistance to life-threatening infections which allows even greater transmission of these infections.

Larger animal behaviour is also being altered by weather changes. Species adapt by moving to areas that allow them to regain the temperatures they were evolved to thrive in. This migration can cause problems such as competition with species already present, lack of food sources, and lack of evolved predators. This species migration is already being witnessed across all of the animal and plant kingdoms. For our own species, according to the Internal Displacement Monitoring Centre [11], one displacement (i.e. an individual being forced to leave their home region) *per second* took place due to extreme weather events in 2023. Conflict over food and water resources is likely to be a feature of climate change. War is catastrophic for all and has a hugely deleterious impact upon the climate.

Air quality is an issue that is increasingly being recognised as a threat to human health. Air pollution is not caused by climate change but is exacerbated by it. Increased atmospheric heat is more likely to trap the particles and gasses at ground level. Ozone and particulate matter are particularly vulnerable to this and increase considerably on warmer days. As we saw previously, wildfires are far more common in our warming world and release a toxic stew of chemicals often over many miles. Wildfires are becoming more common, more northerly, lasting longer and so releasing more and more pollutants.

The World Meteorological Organization has stated that extreme temperatures are not the only hazard from heatwaves; they also cause pollution-related health threats. In their annual air quality and climate bulletin, the meteorologists highlighted a 'vicious cycle' of climate breakdown and air pollution. They found that heatwaves sparked wildfires in the northwestern US and heatwaves accompanied by desert dust intrusions did the same across Europe. The hot temperatures in Europe, which in 2022 were record-breaking, led to much higher levels of particulate matter in the air. [12]

Air pollution has been implicated in a whole range of diseases – from dementia to heart disease to respiratory problems. Large particulate matter is visible to the naked eye and causes its own problems, especially to those with pre-existing lung disease. There is increasing evidence for the harm that tiny invisible particles are doing to our health. These tiny particles (called $PM_{2.5}$) when breathed in, can pass directly into the circulation and then to any part of the body. They lodge in blood vessels and cause chronic inflammation, and this is at least part of the harm they do.

One indirect threat to human health involves the societal upheavals that climate change can cause. Governments will find themselves increasingly stretched by the pressures of climate change. Dealing with water and food supply issues, conflicts with neighbours, and large-scale immigration does not leave much time or money to invest in healthcare. Climate change continually produces these feedback loops which will be increasingly difficult to escape from. Initially, those countries who already struggle to afford healthcare for their populations will be affected, but as the threats increase and costs rise, no healthcare system anywhere in the world will be unaffected.

Another forgotten risk of climate change is the direct damage to healthcare itself with the costs of repair and the greater burden of illness. A cyclone or wildfire or flood can quickly destroy hospitals and health centres in that area. Healthcare workers may themselves be directly affected or may not be able to travel to healthcare facilities. Communities where transport is cut off will soon run out of medicines and equipment. Just when the burden on healthcare is so acute a community's ability to address this need can be severely limited.

In the longer term it has been shown that these major events take some years to recover from. As Longman *et al.* describe, our approach to climate-related events isn't always logical or practical. The concept of 'recovery', in the sense of things returning to their previous pre-event state, isn't always possible. Increasingly, we shouldn't think of these events as one-offs, as they are likely to become more and more frequent. As well as trying to build in practical resilience in healthcare we might also need to begin to gain a mental resilience, meaning not being surprised by these events and developing both practical and psychological mechanisms to deal with the short and long-term aftermath [13].

Our warmer world will inevitably cause us all to change our behaviour. Seeking a cooler environment, we will head indoors, use air conditioning (itself a major contributor to greenhouse gases) and travel by car more often. All of these mechanisms to keep out of the excesses of the heat also make us more static. Avoiding heat inducing exercise, moving less, and using cars all reduce our energy output and this puts us at risk of obesity. Obesity, as many societies are discovering, has a huge impact upon many aspects of an individual's health, not least from the risk of diabetes. Widespread obesity also puts its own pressures upon healthcare systems.

Behavioural changes in hot weather are not just limited to reduced activity. Drowning risks rise as people try to cool off [14]. Alcohol intake increases as some use it to combat thirst. Crime has been shown to increase in hot weather, with causes ranging from there being more open windows to there being more people venturing outside [15]. Mental health can be adversely affected with greater ambient heat – as can insomnia [16]. This once again shows that we can never truly predict the adverse consequences of climate change.

Climate change is taking, and will continue to take, its toll on all of Earth's organisms. As the dominant species on the planet we can mitigate this, but only to a point. Where this point is depends upon where you live, what age you are, and your economic circumstances. Those who argue that Net Zero is just too expensive might want to consider a 2018 study from the US. They looked at 10 climate change events in the US and found that the health-related costs, including hospital admissions, emergency department visits, other medical costs and lost wages, totalled US$10 billion. [17]. If we extrapolate this to the remainder of the US and the rest of the world, the costs of climate change are astronomical.

References

1. Environmental Defense Fund. How Climate Change Makes Hurricanes More Destructive. https://www.edf.org/climate/how-climate-change-makes-hurricanes-more-destructive.
2. Geophysical Fluid Dynamics Laboratory 2024. Global Warming and Hurricanes. https://www.gfdl.noaa.gov/global-warming-and-hurricanes/.
3. Xu R, Yu P, Abramson MJ et al. Wildfires, global climate change, and human health. *The New England Journal of Medicine* 2020;383:2173–2181. https://doi.org/10.1056/NEJMsr2028985.
4. Akdis CA, Nadeau KC. Human and planetary health on fire. *Nature Reviews Immunology* 2022;22:651–652. https://doi.org/10.1038/s41577-022-00776-3.
5. United Nations Environment Programme 2022. Spreading Like Wildfire: The Rising Threat of Extraordinary Landscape Fires. https://www.unep.org/resources/report/spreading-wildfire-rising-threat-extraordinary-landscape-fires?gclid=CjwKCAjwloynBhBbEiwAGY25dGNSn1RhLeMs-s3eixLwp5VDnw8jnEuANabHqEuLO90mAiWpN1hEJBoCVq8QAvD_BwE.
6. United State Environmental Protection Agency Climate Change Impacts. Climate Impacts on Agriculture and Food Supply. https://climatechange.chicago.gov/climate-impacts/climate-impacts-agriculture-and-food-supply.
7. Taub DR, Miller B, Allen H. Effects of elevated CO_2 on the protein concentration of food crops: A meta-analysis. *Global Change Biology* 2008;14(3):565–575. https://doi.org/10.1111/j.1365-2486.2007.01511.x.
8. Zhu C, Kobayashi K, Loladze I et al. Carbon dioxide (CO_2) levels this century will alter the protein, micronutrients, and vitamin content of rice grains with potential health consequences for the poorest rice-dependent countries. *Science Advances* 2018 May 23;4(5):eaaq1012. https://doi.org/10.1126/sciadv.aaq1012.

9. Smith M, Myers S. Impact of anthropogenic CO_2 emissions on global human nutrition. *Nature Climate Change* 2018;8:834–839. https://doi.org/10.1038/s41558-018-0253-3.
10. The Guardian 1 Sep 2023. 'A First in Paris'. City Fumigates for Tiger Mosquitoes as Tropical Pests Spread Bringing Disease. https://www.theguardian.com/world/2023/sep/01/paris-fumigates-city-tiger-mosquitoes-carry-zika-dengue-disease-france.
11. Internal Displacement Monitoring Centre. https://www.internal-displacement.org.
12. World Meteorological Organization 2023. WMO Report Highlights Continuous Advance of Climate Change. https://wmo.int/news/media-centre/wmo-annual-report-highlights-continuous-advance-of-climate-change.
13. Longman J, Patrick R, Bernays S, Charlson F. Three reasons why expecting 'recovery' in the context of the mental health impacts of climate change is problematic. *International Journal of Environmental Research and Public Health.* 2023 May 19;20(10):5882. https://doi.org/10.3390/ijerph20105882.
14. Sindall R, Mecrow T, Queiroga AC et al. Drowning risk and climate change: A state-of-the-art review. *Injury Prevention* 2022;28:185–191. https://injuryprevention.bmj.com/content/28/2/185.
15. Peng J, Zhan Z. Extreme climate and crime: Empirical evidence based on 129 prefecture-level cities in China. *Frontiers in Ecology and Evolution* 2022:10. https://doi.org/10.3389/fevo.2022.1028485.
16. Rifkin D, Long M, Perry M. Climate change and sleep: A systematic review of the literature and conceptual framework. *Sleep Medicine Reviews* 2018;42:3–9. https://doi.org/10.1016/j.smrv.2018.07.007.
17. Limaye V, Max W, Constible J et al. Estimating the health-related costs of 10 climate-sensitive U.S. events during 2012. *Geo Health* 2019;3(9):245–265. https://doi.org/10.1029/2019GH000202.

3 Climate change and eye disease

Climate change brings threats to all aspects of our health but will have a particular impact on those parts of our body that directly interface with the world. This of course includes our eyes. As we will see, climate change will have specific effects upon our eyes, but at the same time, any of the risks to our general health can inevitably impact our eyes and visual system.

The eyes are as susceptible to changes in our body as any other organ. They have a rich blood supply that makes them sensitive to changes in haemodynamics. Most systemic diseases have some ocular effects and this, combined with their external position, puts our eyes in the vanguard of risk in a changing climate.

This chapter brings together the current evidence for these threats to our sight and the consequences if our climate continues to change.

3.1 Specific effects of climate change on the eye

The eyelids

Our skin is the organ most likely to be damaged by climate change. The skin around the eyes and on the eyelids is prominent and usually unprotected, making it particularly susceptible to the Sun's rays.

Skin cancer is the world's most common cancer [1] and the incidence has increased over the latter part of the 20th century. Over 126,000 deaths due to skin cancer occurred in 2018. 5–10% of all skin cancers occur on the eyelids [2].

Both melanoma cancers and the non-melanoma basal cell (BCC) and squamous cell carcinomas (SCC) are becoming more common. The rate of melanomas has increased in the US 3-fold over the last 40 years. The incidence of BCC and SCC has doubled in Scandinavia since 1960 and increased 4-fold in Australia [1].

Climate change has helped drive the increased number of cases. Ultraviolet radiation is by far the main cause of skin cancer with a direct mutagenic effect on DNA. The loss of the stratospheric ozone layer in the last century from our

DOI: 10.4324/9781003512608-4

use of CFC refrigerants is most likely responsible for the late 20th century increase in skin cancers [1].

The slow recovery of the ozone layer does hold some hope of future reductions in skin cancer incidence. The climate is complex though – so complex that we still do not fully understand it. Other changes we are making to the atmosphere via the burning of fossil fuels have an impact upon the ozone layer and can continue to deplete it. There is evidence that increased atmospheric temperatures reduce relative humidity. This means that less UV radiation (UVR) is filtered by atmospheric water molecules and so more hits the Earth and its occupants [3]. Added to this it has been shown that air pollution also enhances UVR penetration [4].

These are not the only effects that these environmental pressures have on the skin. Increased heat has been shown experimentally to enhance the ability of UV radiation to promote cancerous changes in cells [5]. Our warming environment could therefore be having a direct influence on skin cancer risks as well as on the amount of UVR. There is also a behavioural effect: as cold climes become warmer and seasonal patterns change, people are more likely to go outside and increase their overall sun exposure. A specific risk of skin cancer is previous episodes of sunburn and in our warmer world these episodes are likely to increase [6].

Finally, air pollution has been implicated in the rise of skin cancers. Topical polycyclic aromatic hydrocarbons have been shown experimentally to induce formation of squamous cell carcinomas. Similarly, they demonstrate a synergistic effect with ultraviolet A via oxidative DNA damage [7]. Thus, as with UVR, air pollution both increases the amount of UVR the skin is exposed to and can worsen the effects of that exposure.

As we can see from the above, there are a number of interacting factors that explain the increases in skin cancers and that these risk factors will only increase with climate change. We will see in the next chapter the current prevalence of eyelid cancers globally. As they represent 5–10% of all skin cancers, and as the eyelids are such a small area compared to the overall skin area, they are a relatively high risk site for these cancers [8].

As well as malignancies, climate change is likely to increase the rate of eyelid contact dermatitis. This is an inflammatory reaction involving the eyelid skin that is caused by contact with a trigger substance. It may be due to allergy (allergic contact dermatitis) or irritation (irritant contact dermatitis). Eyelid dermatitis is also called eyelid eczema [9].

Air pollution has also been shown to exacerbate atopic dermatitis and other inflammatory skin conditions [10]. The eyelids are often affected by this as the thin skin of the eyelids is particularly sensitive to irritants and allergens and is thus prone to develop contact dermatitis. There is also evidence, as we will see in the next section, that the allergens themselves are changing under the influence of climate change.

The conjunctiva

The eye is one of the two organs (the other being the skin) directly susceptible to the Sun's rays. This consists of radiation both from direct sunlight and sky scattering and also from reflections from clouds, the ground, and any other reflective surfaces such as ice and snow. There is an increasing body of epidemiological evidence associating this radiation with the development of photochemical damage to ocular tissues. This damage can affect any part of the eye but especially those structures that receive most solar radiation. The first of these is the conjunctiva which covers the surface of the eyeball [11].

A healthy conjunctiva is essential for the healthy functioning of the external eye and thus for its ability to remain optically clear. As with the eyelids, the conjunctiva creates a barrier to the external world to protect the clarity of vision through the cornea. As with the eyelids, it is directly exposed to the external world.

Inflammation of the conjunctiva – conjunctivitis – is a common condition caused by infection, inflammation, or allergy. We will deal with the more serious infections of the conjunctiva, such as trachoma, in the infectious diseases section.

Allergic conjunctivitis is likely to increase significantly with the pressures of climate change. Rising temperatures and frost-free winters are allowing longer pollen seasons in many plants. The increase in atmospheric CO_2 and temperature means more plants are producing more pollen in the spring. There is also evidence that some pollens are becoming increasingly allergenic [12].

Separately from allergy (though likely synergistic), air pollution seems to have a direct effect upon the conjunctiva. One study found that amongst air pollutants, carbon monoxide was most highly associated with the prevalence of conjunctivitis. Though less of an association was found, there was still an increase in prevalence with sulphur dioxide, PM_{10}, nitrogen dioxide and ozone levels [13]. Aik *et al.* found that every 10µg/m3 increase in $PM_{2.5}$ was associated with a 3.8% cumulative increase in the risk of conjunctivitis over the following week [14].

Neoplasms of the conjunctiva are rare, but there is evidence that conjunctival melanomas are becoming more common [15]. It has also been shown that these melanomas are very similar to eyelid melanomas and so likely to be associated with UV radiation. Thus the environmental risk factors for skin melanoma we saw for the skin should hold for conjunctival tumours. This of course means that they will become more common with climate change [16].

Flood events have been shown to increase the risk of infectious conjunctivitis of various types. The reasons for this are a combination of pathogen growth (bacteria and viral) in stagnant or polluted water, decreased capacity for personal hygiene, such as face and hand washing and populations being forced together in smaller spaces with the increased risk of a direct infection spread [17].

Dry eye

Dry eye has numerous causes, from reduced tear production to enhanced tear evaporation, but essentially means tears are not adequately lubricating the surface of the eye. The results of dry eye can range from minor inconvenience to sight-threatening.

Dry eye disease (DED) already represents a significant burden to sufferers, to healthcare providers, and to economies. We will discuss this more fully in the next chapter but in epidemiological studies performed globally, the prevalence of dry eye disease (DED) ranges from 5 to 50% of the population [18]. This prevalence has been shown to be increasing [19].

DED is associated with a number of systemic diseases but environmental factors also have a negative impact. Associations between raised environmental temperature and diagnosis of DED have been found [20]. A link between temperature variations has been found to be more significant for dry eye than absolute temperature [21]. Other studies have not found this association and, as DED is multifactorial, it is hard to isolate temperature as a direct cause. Allergies in spring or the drying effect of indoor heating in the winter are examples of some confounders. It is worth noting that both air conditioning and heaters cause significant drying of the indoor atmosphere and so increase the risk of dry eye. In a warmer world where we are increasingly forced into controlled temperature environments it will be important to remember this [22].

There is good evidence that air pollution worsens dry eye disease. A Chinese study, [23] found that same day exposure of $PM_{2.5}$ (1.02, 95% CI = 1.01–1.03, $p < 0.01$) and PM_{10} (1.01, 95% CI = 1.003–1.02, $p < 0.01$) increased the risk of outpatient dry eye visits. Additionally, this exposure effect on dry eye disease lasted for up to three days. In another Chinese study, based upon a clinic population [24], extreme air pollution levels for >143 days in a year increased the risk of a dry eye diagnosis by 2.01 (95% CI = 1.79–2.25; $p < 0.0001$). Although not all studies have shown this effect, there is a general trend for outdoor pollution to be associated with dry eye disease [25].

The main pollutants that worsen dry eye disease have been identified. A Taiwanese study [26] found significant associations between nitrous dioxide (OR 1.07; 95% CI = 1.04–1.10; $p < 0.001$) and carbon monoxide (OR 1.12; 95% CI = 1.03–1.21; $p < 0.01$) and dry eye diagnoses. Every 10 parts per billion increase in atmospheric nitrous dioxide was associated with a 6.8–7.5% increase in dry eye diagnoses and every 1 part per million increase in carbon monoxide was associated with a 10.5–11.6% increase in dry eye diagnoses.

The Chinese study mentioned earlier [23] found an association with nitrous dioxide (1.03; 95% CI = 1.01–1.05; $p < 0.05$), sulphur dioxide (1.07; 95% CI = 1.02–1.11; $p < 0.01$) and carbon monoxide (1.0; 95% CI = 1.0–1.0; $p < 0.01$) levels and dry eye disease. This study did not show a link with ozone pollution and dry eye disease but other studies have [27]. A South Korean population found that a 0.003 parts per million increase in ozone level was associated with both an increased diagnosis and worsening symptoms of dry eye (OR 1.31; 95% CI = 1.12–1.53; $p = 0.002$).

The causes of the increasing prevalence of dry eye seem to be multi-factorial, though the ultimate mechanism of damage is probably the same. Oxidative stress and inflammation caused by pollution, heat or UV radiation are the most likely mechanisms [28]. These factors will all be worsened as our climate changes.

The cornea

The cornea represents the window to the world of the eye and visual system. Its clarity is vital for our sight but again suffers the hazard of directly interfacing with the environment. We have seen in the sections above the crucial nature of eyelid function, conjunctival health, and tear film stability on ocular health. All these are working to keep the cornea optically clear and each one, when it dysfunctions, can mean loss of this corneal clarity.

Any inflammation of the lids and conjunctiva can affect the health of the cornea. Severe dry eye will significantly harm the cornea, as well as being miserable for the sufferer. The factors that worsen dry eye that we discussed in the previous section will also affect the cornea. Severe dry eye is a significant risk to the cornea and directly affects its clarity. Similarly, severe allergic eye disease presents a risk to this clarity, as can a number of infective diseases.

Ultraviolet radiation impinges directly upon the cornea when it hits the eye – the structures affected are determined by the wavelength. Shorter wavelengths are more biologically active and absorbed mainly by the cornea. The cornea therefore acts as a very important barrier to a significant proportion of UVB (280–315nm) and prevents it from reaching the posterior structures of the eye. UVR at 300 nm has been shown to cause cell apoptosis in all three layers of the cornea [29]. UVA (315–400nm) can penetrate the full thickness of the cornea, even reaching the anterior portion of the lens. This absorption of energy by the cornea comes at a cost and epidemiological data indicates that UVR is a contributing factor for a multitude of diseases of the cornea [30].

These include a progressive scarring of the cornea called pterygium. This is caused by dry eye and UV radiation so is much more common in outdoor workers and those in hotter, drier climates. Pterygium is a common finding in those circumstances and is found in around 12% of the population globally [31]. In most cases the pterygium does not affect vision, but it can sometimes cause a change in the shape of the cornea (astigmatism) and, if severe, can cross the visual axis with a resulting drop in vision. A study from Myanmar found that pterygium was associated with 0.4% (95% CI 0.04 to 1.3) of binocular visual impairment and 1.0% (95% CI 0.6 to 1.8) of visual impairment in a least one eye [32].

When a pterygium becomes sight-threatening surgical intervention is the only treatment. It is highly likely the incidence of pterygia will increase in number with climate change and this will create an increased surgical burden – often in parts of the world where surgical facilities are not available to all.

UV radiation has been linked to other corneal pathologies, including keratoconus and Fuchs endothelial dystrophy and various types of corneal inflammation (keratitis) [28]. Keratoconus is also closely related to allergic eye disease – which, as we saw previously, is itself projected to worsen as the globe warms. Keratoconus and Fuchs dystrophy can lead to progressive, significant visual impairment. Though conservative measures may be effective, a significant number of patients with this condition will need quite major ocular surgery.

Corneal surface neoplasia can occur and is associated with a number of factors but UV exposure and pollution do seem to be risk factors [33]. Although rare, this ocular surface neoplasia is the most common non-pigmented tumour of the ocular surface with a worldwide age-standardised rate of 0.26 cases per 100,000 per year. There is a higher incidence in African countries of 3–3.4 per 100,000/year and it is becoming more common worldwide [34].

The lens

As with the cornea, the lens has to retain optical clarity or vision will be degraded or lost. Unlike the cornea the lens sits within the eye so is protected from the external environment. What it is not fully protected from is some wavelengths of UV radiation [28]. This radiation, as it passes through the lens, imparts energy and this energy within the lens is responsible for damage to the lens proteins. Thus the longer wavelengths that reach the lens contribute to cataract formation [35]. There have now been a number of studies that confirm this relationship between UV radiation and cataract formation [36].

There are a number of other environmental factors that can cause or worsen cataracts. A Chinese study indicated that higher ambient temperatures were associated with higher reported cases of cataract [37]. Firm associations have been made between those who work in warm environments, such as the glass and metal industries, and cataract formation. Experimental work also seems to show a direct connection between ambient heat and clouding of the natural lens [38].

Dehydration has also been shown to be a risk factor for cataract. A case-control study from India found that an estimated 38% of blinding cataract was attributable to repeated bouts of dehydration from severe diarrhoeal disease and/or heatstroke. This risk of blinding cataract was strongly related to the level of exposure to dehydrational crises in a consistent and dose dependent manner [39]. As discussed in Chapter 1, we know that floods and droughts and their associated diarrhoeal diseases will be more prevalent and severe as the climate warms.

Yet again air pollution has been implicated and associations have been found with an increased risk of cataract formation. One study showed that long-term exposure to $PM_{2.5}$ was associated with age-related cataracts. A dose response relationship was found in this study, with higher exposure levels being more likely to be associated with cataract reporting [40]. A UK study confirmed this association and found that higher exposure to $PM_{2.5}$ was associated with a 5% increased risk of incident cataract surgery per interquartile range [IQR] increase. Compared to the lowest quartile, participants with exposures to $PM_{2.5}$, nitrous dioxide and nitrous oxide in the highest quartile were 14%, 11% and 9% more likely to undergo cataract surgery, respectively. A continuous exposure–response relationship was observed, with the likelihood of undergoing cataract surgery being progressively higher with greater levels of $PM_{2.5}$, nitrous dioxide and nitrous oxide (P for trend $P < 0.001$). [41].

It might be argued that cataracts are not a serious consequence of climate change as they can readily be removed. However, this is still a surgical procedure with its attendant risks. Not everyone can afford or has access to cataract surgery and we will discuss the huge burden of cataract blindness in the forthcoming chapters.

Glaucoma

Glaucoma occurs in more than 2% of the population over 60 and can result in devastating blindness. It is a disease of the optic nerve with the main risk factor being raised pressure inside the eye. There is however, increasing evidence that external factors can also raise the risk of glaucoma.

Wang *et al.* found that, amongst other factors, higher ambient ultraviolet radiation and a higher level of air pollution were significantly associated with a higher burden of glaucoma [42]. Pasquale *et al.* reported an association between sun exposure and the risk of pseudoexfoliative glaucoma [43]. A link has been reported between the blood supply of the retina and optic nerve and air pollution and this could be the mechanism for the damage in glaucoma [44].

A study using the UK Biobank found that participants resident in areas with higher $PM_{2.5}$ concentration were more likely to report a diagnosis of glaucoma (OR 1.06, 95% confidence interval [CI] = 1.01–1.12, per interquartile range increase $P = 0.02$). Higher $PM_{2.5}$ concentration was also associated with a thinner nerve fibre layer in the retina – which can be a sign of glaucoma. They did not find an association between $PM_{2.5}$ concentration and intraocular pressure suggesting a directly toxic effect on the nerve cells or their blood supply [45].

Similarly, Sun *et al.* found that increased $PM_{2.5}$ exposure was an independent risk factor for glaucoma (ORs > 1, $P < 0.05$) [46]. Min *et al.* found that short-term and long-term exposure to PM_{10} was associated with an increased incidence of childhood glaucoma [47].

The retina

The retina, despite evolution carefully protecting it from the outside world, will also be affected by climate change. There are a huge number of retinal diseases and conditions and increasingly they can be connected to changes in the external environment.

One study found that retinal detachments were more frequent with raised ambient temperatures [48]. Another study found that 38.2% retinal detachments occurred in the hot humid summer months (P = 0.046) but only 22.4% in the cool winter months. The peak minimum temperature, peak maximum temperature and sun hours were related to this finding [49]. Kim *et al.* reported that there was a significant positive correlation between the monthly average of the daily temperature range and the number of rhegmatogenous retinal detachment operations [50]. As this is a condition that needs complex intervention by specifically trained surgeons, any increase in incidence would require significant resources and expertise. This may be a particular problem in parts of the world without adequate coverage of retinal surgeons or equipment.

Age-Related Macular Degeneration (ARMD) is the leading cause of irreversible visual loss in those over 65 years of age. The only definite risk factors are a family history and increasing age, however, as with glaucoma, there is increasing evidence that environmental factors may have an impact.

Air pollution is again one of the leading candidates for this. A Taiwanese study found that chronic exposure to the highest quartile of ambient nitrous dioxide or carbon monoxide significantly increased the risk for ARMD [51]. Grant *et al.* found that In single-pollutant models, higher values of $PM_{2.5}$ were associated with an increased odds of visually impairing ARMD [52]. Millen *et al.* reviewed all the current evidence linking air pollution with eye disease [53]. One study they found reported an increased odds of self-reported visually impairing ARMD with an increase of one IQR in $PM_{2.5}$. The UK Biobank study found increased odds of self-reported ARMD with continuous increasing concentrations (μg/m^3) of $PM_{2.5}$. Another study found increased odds of early ARMD with exposure to PM_{10} in the prior 3–6 years. Increased odds of early ARMD was observed for high versus low exposure to sulphur dioxide (SO_2) in the prior 1–2 years.[53].

It has been suggested that the mechanism for this is that air pollution induces oxidative stress which activates inflammatory pathways and increases coagulation, thus damaging the retina. The retina has a high consumption of oxygen thus making it very susceptible to this oxidative stress. This is especially so in the central retina -the macula [54].

Other types of retinal pathologies have been linked with air pollution. One study from Taiwan demonstrated a positive association between diabetic retinopathy (DR) and PM ≤ 2.5 and 2.5 to 10μm in diameter, with odds ratios of 1.29 (1.11–1.50) and 1.37 (1.17–1.61) respectively. In a national cross-sectional study in rural China, Shan *et al.* found that increased exposure to a high concentration of $PM_{2.5}$ was related to an increased risk of diabetic retinopathy among diabetic patients in rural China [55].

Perhaps unsurprisingly, greater exposure to UV radiation has also been linked to a greater risk of ARMD. Although much of the UV is absorbed during the journey to the retina, some still impinges upon it. It seems that even this limited amount can do damage and induce significant oxidative stress in the retinal pigment epithelium [56]. Schick *et al.* found an association between greater sunlight exposure during working life and a higher prevalence of ARMD [57]. Similarly, Delcourt *et al.* found in a population based study that lifetime ambient UVR exposure gave a higher risk of ARMD development [58].

Malnutrition in all its forms can have a serious impact on retinal function. We will discuss this in detail later in this chapter when we look at nutrition and the eye. Similarly, we will look at cardiovascular and circulatory diseases in a later section as they can have a devastating impact upon the retina. For example, Cheng *et al.* looked at central retinal artery occlusion (CRAO) and air pollution. CRAO is a rapid blockage of blood supply to the retina and results in death of the retinal cells within a few hours. They found that the risk of CRAO onset was significantly increased (OR, 1.09; 95% CI, 1.01–1.17; P = 0.03) during a 5-day period following a 1 part per billion increase in NO_2 levels. The risk of CRAO onset also significantly increased in patients with hypertension and in patients over 65 years old, after 1 day of elevated SO_2 levels (OR, 1.88; 95% CI, 1.07–3.29; P = 0.03 and OR, 1.90; 95% CI, 1.13–3.21; P = 0.02, respectively) [59].

The uveal tract

The uveal tract consists of the iris, ciliary body, and choroid. Whilst an integral part of the eye, they also have their own pathology and, once again, evidence indicates that climate change can adversely affect this.

Inflammation of the uveal tract – uveitis – has been associated with increased air pollution. A Chinese study found that a 10μg/m3 increase in $PM_{2.5}$ concentration was associated with a one case per 10 individuals increase in uveitis onset. They concluded that $PM_{2.5}$ concentrations above a specific level were responsible for 13% of novel uveitis cases [60]. The same group also found an association between higher ambient temperature and uveitis onset. A 1°C increase in monthly temperature was associated with a small increased risk of uveitis [61]. Uveitis can vary from the mild and self-limiting to the severe and sight-threatening. It often requires the use of topical and sometimes systemic steroids which in themselves can be a risk to sight and general health.

Ocular melanomas develop in the uveal tissue. Though they are rare they are the commonest primary cancer of the eye. There is some evidence that increased UV exposure is a risk factor for this melanoma [62]. A Swedish study showed a lower incidence of uveal melanoma with increasing latitude [63]. Yu *et al.* produced similar findings and suggested it was due to decreased sun exposure further north [64].

3.2 General effects of climate change on ocular health

Trauma

As we saw in the previous chapter, the energy that is currently being put into our climate system has to be discharged somewhere. This means stronger winds, heavier rain, more wildfires, higher tides, and heatwaves. Whilst these are already taking their toll on our general health, our visual system is equally threatened.

Wind events can cause a whole range of damage and the eyes being relatively exposed can sustain serious injury. Wind-blown objects can cause blunt injury to the orbits or damage the eyes directly – these range from minor foreign bodies and corneal abrasions to sight-threatening injury. Sharp objects propelled with great force can cause penetrating eye injuries – which are almost invariably sight-threatening.

Those trying to escape or prepare for wind events have a higher rate of car accidents. Windscreen trauma to the eyes can occur in these circumstances. Broken glass and flying glass can cause devastating damage to an eye.

A study showed that, following Hurricane Harvey, most ophthalmic injuries were corneal/anterior segment injuries (38–46% of patients) and vitreoretinal injuries (from 17–23%) [65]. A different study from the same hurricane noted an increase in ocular infections most likely caused by the flooding that followed the event [66].

Flash flooding contains a huge amount of power and so can significantly damage any part of the body – including the eyes and orbit. Objects can be carried at great speed and have serious impacts. The more chronic risks from flooding are dealt with later.

Wildfires present numerous hazards to the eye. In the acute situation thermal burns of the eyelids and cornea are possible. These need rapid attention. Even those at some distance from the fire can be affected with large airborne particles causing irritation and damage to the ocular surface [67].

It is important to remember that primary care for these injuries may not always be available. Damage to healthcare facilities from tornadoes, flooding or fire may put them out of action and healthcare workers may have been injured or be unable to travel to healthcare sites. Similarly, external help may not be able to arrive in a timely manner due to infrastructure damage. Minor or reversible injuries may not get the treatment needed, with the risk of progression to more serious damage.

It is also important to remember that this damage to healthcare facilities can take significant amounts of time to recover. This can have effects on chronic illnesses, such as glaucoma and macular degeneration. Following Hurricane Katrina in 2005 much of the New Orleans healthcare system was affected. More than a dozen hospitals were damaged and thousands of doctors dislocated, with virtually all New Orleanians losing access to their usual healthcare providers. Individuals with acute or chronic conditions were particularly hard hit. According to US government officials, 2,500 hospital patients in Orleans Parish alone were evacuated. In addition, renal dialysis centres across Louisiana with caseloads of between 3,000 and 3,500 patients were destroyed and only half of their patients were accounted for several weeks after the storm hit [68].

Infectious diseases

Of all the threats that climate change brings to human health, infectious diseases are likely to be one of the greatest. Climate change will alter the distribution, type, and severity of infectious diseases both in ways we can predict and in ways we cannot. Many micro-organisms will flourish in the warmer world; those who don't will migrate to cooler areas. Vectors will also change, and floods and droughts will create perfect environments for many diseases.

In this section we will look at the major ocular infections that are likely to be worsened by climate changes. Many systemic infections can also have an ocular effect but we will look at those that are primarily ophthalmic. It is important to remember that many of these are already endemic in some parts of the world, causing untold misery. That this will worsen is bad enough but without doubt, under the pressure of climate change, these diseases will spread further and further into areas they have never before been. Developed countries are suffering the effects of the changing climate and this is so with eye diseases. The Southern US and Southern Europe are already reporting eye diseases that were thought only to occur nearer the equator.

Trachoma

In addition to rainfall and land cover, studies have reported associations between active trachoma and distance to water source [69], temperature [70], altitude [71], markers of socio-economic status [72] and markers of personal or household hygiene, such as facial cleanliness [73].

Trachoma is a leading cause of blindness in sub-Saharan Africa. It is caused by bacterium *Chlamydia trachomatis* which is spread from person to person by direct contact with secretions or by flies moving from an infected person. It causes inflammation and then scarring of the conjunctiva. If this scarring progresses, the eyelids are turned inwards and this, combined with the conjunctival changes, causes corneal scarring. Untreated, this scarring can be so severe the sufferer becomes blind. It is a public health problem in 42 countries and is responsible for the blindness or visual impairment of about 1.9 million people [74]. We will discuss the burden of this devastating disease in the next chapter.

Climate change brings the ideal conditions for the spread of trachoma. Flooding and drought mean that personal hygiene especially face and hand washing is reduced. This allows optimal spread of the bacteria between individuals. Stagnant water allows huge colonies of flies to breed and similarly spread the infection. Higher temperatures also enhance fly numbers, whilst conversely higher altitudes decrease the risk of trachoma [75]. As we have seen, one effect of climate change is to push more people into smaller areas – either for a safer temperature or water or food. This again creates an ideal opportunity for the infection to spread – both from human proximity and stresses on water supply.

The World Health Organization (WHO) describes something called Emerging Infectious Diseases (EIDs). These include new unknown diseases which are first diagnosed and old diseases which have evolved or mutated from already existing agents and gained new features. The latter include adaptation to new hosts or target populations, new geographic distribution, a new clinical picture, a new epidemiological profile, new spread pattern or increased resistance to therapeutics. Climate change will be a powerful driver of this [76]. Amongst the list of diseases that fit this definition the WHO have included trachoma. This means it is one of the likeliest diseases to spread to countries where it has not previously been found.

Onchocerciasis

Another infectious disease on the WHO list of EIDs is Onchocerciasis, also called River Blindness. It is the second commonest infectious blinding disease in the world. Onchocerciasis is a tropical, parasitic disease caused by the filarial worm *Onchocerca volvulus*, transmitted by repeated bites of infected blackflies. These flies breed along fast-flowing rivers and streams and when a female blackfly bites an infected person during a blood meal, it also ingests microfilariae which develop further in the blackfly and are then transmitted to the next human host during subsequent bites. Once the microfilariae enter the bloodstream they move to the skin and then to the eyes [77].

The presence of the microfilaria in the eye causes an inflammatory response and this is what damages the eye. Any part of the eye can be involved – from keratitis to retinal involvement to optic neuropathy [78]. This can lead to chronic inflammation of these ocular structures and can ultimately lead to visual impairment and blindness.

Climate change will increase the prevalence and increase the geographical spread of onchocerciasis [79]. This is mainly due to the fly vector, which has been shown to proliferate at 3–7°C above the usual temperature for that area. Poor sanitation following droughts or floods or natural disasters also allows greater fly infections. Seasonality has an influence on fly numbers and behaviours and longer warm seasons will give greater opportunities for infection.

Toxoplasmosis

Toxoplasma gondii is an intracellular parasite and is widespread in the natural world. Approximately one-third of humans globally are estimated to be chronically infected with *T. gondii*, with certain geographic regions having a far higher rate of infections [80].

For many years, ocular toxoplasmosis was considered to be the result of recurrence of the congenital form of the disease directly passed through the placenta. However, more recent reports support the view that acquired infections might be a more important cause of ocular diseases than congenital ones. Whichever form it is, ocular toxoplasmosis is a progressive and recurring necrotising retinitis, with vision-threatening complications such as retinal detachment, choroidal neovascularisation, and glaucoma, which may occur at any time during the clinical course. It is the main cause of posterior uveitis in the US and Europe [81].

Changes in the climate are likely to increase the risk of ocular toxoplasmosis. Increased temperature and humidity have been shown to increase the survival and infectivity of the cysts. Flooding and heavy precipitation can wash the cysts into the water supply especially with poor soil structure following droughts. Climate changes can also affect the vectors for the diseases – both the vector's life cycle itself and their geographical spread [82]. As for most infectious diseases, human activities such as deforestation, urbanisation, and agricultural practices that result in habitat loss, reduce biological diversity, and so provide favourable conditions for the occurrence and spread of parasitic zoonosis. Climate change itself can do this and acts synergistically with this habitat loss [83].

A good example of the above was a large outbreak of Toxoplasmosis reported in French Guiana, which occurred after an unusual flood combined with warm temperatures [84]. Floods and warmer temperatures are, of course, the conditions we will increasingly expect in our warmer world.

Acanthamoeba

Acanthamoeba is a protozoa that exists all over the world in water, air, soil, and dust. In normal circumstances it rarely causes humans any problems but has become an increasing issue secondary to contact lens wear. Poor care of contact lenses, overuse, washing in tap water or swimming or showering with contact lenses in place can allow the acanthamoeba to adhere to the contact lens and infect the cornea. Infection slowly progresses and can, if untreated, cause significant scarring of the whole cornea.

The incidence of acanthamoeba infections seems to be increasing [85]. Though the reasons for this aren't clear, the environments we will increasingly see with climate change will be increasingly hospitable to acanthamoeba. Acanthamoeba numbers in fresh and marine water increase with the water temperature. They can survive droughts and therefore become more concentrated in diminishing lakes and rivers. They can survive in a wide range of water conditions, from fresh to high salinity. Weak soil structure from droughts or wildfires means that the acanthamoeba can more easily be washed into the water supply. Finally, one of our behavioural responses to increasing ambient heat is to bathe or swim in cooler water and this increases the opportunities for infection [86].

Fungi

Of the many disease threats in our future warming world fungal infections are likely to be high on the list. Though fungi are everywhere they generally do not pose a threat to our lives, unless through opportunistic infections in those with compromised immune systems. However, as global temperatures rise so does the incidence of fungal infections. Only a small percentage of the estimated millions of fungi on Earth can infect people. Currently most fungi cannot survive at human body temperatures (around 98.6 degrees Fahrenheit) and need cooler environments. With rising temperatures, fungi may be evolving to live in these warmer conditions – including the human body. New fungal diseases may therefore emerge as fungi become more adapted to surviving in humans. Heat may also cause other genetic changes that can increase the ability of fungi to infect people [87].

Fungal infection of the eye can occur anywhere from the cornea to the retina. Corneal infections come from external sources by direct inoculation or secondary to an abrasion. Intraocular fungal infection can occur by penetration of the globe either by direct trauma or smaller sharper objects entering the vitreous. More often, intraocular infection is via endogenous infection (fungi carried by the bloodstream). Whatever the route or area of infection they characteristically are slow growing, difficult to diagnose early, and equally difficult to treat. Fungal keratitis alone affects over half a million people per year globally [88].

Evidence indicates that fungal keratitis is becoming more common. It is linked to higher ambient temperature, higher humidity, increased rainfall, and heatwaves. Similarly, other fungal infections have been linked to high temperatures, increased rainfall and during heat waves [28, 89]. Severe weather events, such as dust storms, tornadoes, and wildfires, cause environmental disruptions that can spread fungal spores into the air and potentially transport them to new locations. Meanwhile those new locations are becoming more hospitable for the fungus as temperature and precipitation patterns change [90].

Other infections

We have discussed some of the main ocular infections but there are many more and it is worth repeating that most systemic infections (of any organism) can have an ocular effect. It is also worth repeating that few infectious diseases will not become more of a threat as we change our environment. The WHO report on *Health and Climate Change* [91] is very comprehensive. In the infectious diseases section many emerging infectious diseases (as well as the ones we have already discussed) can have ocular components. Examples include dengue fever, buruli ulcer (*Mycobacterium ulcerans* infection), yaws, treponematoses, leprosy (bacterial diseases), Chagas disease, trypanosomiasis, leishmaniasis, dracunculiasis, cysticercosis, echinococcosis, Chikungunya, Zika virus, West Nile virus and yellow fever [92–94].

According to the WHO, 20 core so-called Neglected Tropical Diseases are the most likely candidates to re-emerge in the developed countries due to global warming. (The WHO describes Neglected Tropical Diseases (NTDs) as a diverse group of infections that are mainly prevalent in tropical areas. They are 'neglected' because they are almost absent from the global health agenda. NTDs tend to have very limited resources as they are almost ignored by global funding agencies [95]).These are dengue fever and rabies (viral diseases), buruli ulcer (*Mycobacterium ulcerans* infection), trachoma (*Chlamydia trachomatis*), yaws (spirochete), leprosy (bacterial diseases), Chagas disease, trypanosomiasis, leishmaniasis, dracunculiasis, cysticercosis, echinococcosis, foodborne trematodiases, lymphatic filariasis, and onchocerciasis. All of these can affect the eye and vision [96].

Finally in this section, the mainstay of our protection from these pathogens – antibiotics – may be threatened by climate change. Research has shown that increased temperatures increase both the rate of bacterial growth and the rate of the spread of antibiotic resistant genes between micro-organisms. There is also evidence that increased particulate matter, particularly the very small particles, increases antibiotic resistance. Data from 2013–2015 suggest that an increase of the daily minimum temperature by 10°C will lead to an increase in antibiotic resistance rates of *Escherichia coli*,

Klebsiella pneumoniae and *Staphylococcus aureus* bacteria by 2–4% (up to 10% for certain antibiotics) [97].

Cardiovascular disease

The eye has a rich blood supply and any interruptions to this can have serious effects on vision. Similarly, the large portion of the brain that is devoted to sight has a generous blood supply and interruption to this is potentially very serious.

The eye and ocular structures are supplied by the carotids, thus any disruptions to carotid flow or emboli can impinge upon the eye. The eyes' own blood vessels can have their own problems – the central retinal artery can become blocked and this can have devastating effects upon the retina (see the Retinal section above). The central vein can also become blocked, with equally disastrous consequences for the eye. Even smaller vessel blockage, depending upon the site, can irreversibly affect the vision. The risk factors for all these – high blood pressure, cholesterol, and especially diabetes – are the same as for the rest of the vascular system.

Well over 50% of our brains are involved in visual processing, thus anything that damages our brains has a high chance of damaging our visual function. Strokes, bleeds, emboli, trauma or infection can either damage the visual cortex itself or the visual pathways or the parts of the brain that support the visual system. One study found that nearly three-quarters (752/1033) of stroke sufferers had visual problems. 56% had impaired central vision, 40% eye movement abnormalities, 28% visual field loss, 27% visual inattention and 5% visual perceptual disorders [98].

Thus cardiovascular disease can have very significant effects upon the eye and the visual centres of the brain. Cardiovascular disease is very likely to be worsened by climate change, in a number of ways.

Chronically increased ambient heat has been shown to increase cardiovascular events in populations. Acute exposure to heat in the form of heatwaves has long been known to trigger heart attacks and strokes. This is especially so with high humidity as, in an attempt to cool down, blood is diverted to the skin so that heat can be dissipated into the air. This means that the heart has to work much harder to pump blood around the rest of the body and this increased work by the heart is what can precipitate a heart attack. In very hot weather more blood is diverted to the skin and this makes the blood 'stickier' and so more likely to clot. Dehydration from the heat also has the same effect on the blood. Blood that is more likely to clot makes heart attacks more likely, as well as strokes, deep vein thrombosis (DVT) and lung emboli [99].

Air pollution has been shown to adversely affect the cardiovascular system. Fine particulate matter ($PM_{2.5}$) has been shown to increase inflammation in the blood vessels and to create harmful chemicals in the bloodstream [100]. The inflammation damages the blood vessels, so clots are more likely to form.

The blood itself becomes more viscous, making it more likely to clot. The changes also increase blood pressure over the longer term. All these combine to significantly increase the risk of heart attacks and strokes and other vascular events. We saw previously that even the tiniest of vessels – of which they are many in the eye and optic nerve – can be directly affected by particulate air pollution [44].

Nutrition

Climate change is already having an impact upon our food supply, as we discussed in an earlier section. Heatwaves, floods, droughts, and wildfires rapidly destroy crops and livestock. Disruption of food chains in the warming oceans reduces a huge source of protein for over three billion of the world's population [101].

Malnutrition is an umbrella term that covers starvation and obesity and also includes diets poor in vitamins and minerals. Whilst poor harvests secondary to droughts or floods most obviously lead to starvation, it has also been shown that heat and drought stressed plants contain fewer nutrients. A number of eye diseases are caused by poor nutrition and most vitamin and mineral deficiencies will have an adverse effect on our eyes [102]. At present these are usually seen in developing countries, often after wars or severe weather events. As our planet warms this is likely to worsen and increasingly other parts of the world will begin to have cases. A balanced diet is as important for your eyes as for any other organ but there are some specific nutritional eye diseases which are the most common globally and so are likely to be the ones most affected by climate change.

Vitamin A deficiency

Vitamin A deficiency is already the leading cause of preventable blindness in children worldwide. An estimated 250,000 to 500,000 children become blind every year because of vitamin A deficiency. Half of these children die within a year of losing their sight [103]. Deficiency is due to poor intake of vegetables and fish and/or because of poor intestinal absorption. The latter is seen in a number of infectious diarrhoeal diseases [104]. Measles infection can also reduce vitamin A levels very significantly [105].

Vitamin A deficiency has a profound effect on the eyes and vision. This is called Xerophthalmia. Anteriorly it ranges from mild dry eye to severe corneal damage and scarring and potentially necrosis (keratomalacia) and perforation of the cornea. The latter is not consistent with retention of vision. Deficiency can also affect the retina with the initial symptom being difficulty in night vision and if the deficiency persist this will continue to worsen. Eventually, day vision is also affected and this, unless rapidly reversed, will result in permanent blindness [106].

Vitamin A deficiency is a very good illustration of the hazards of climate change. Climate change leads to poor crop yields and kills livestock and fish. Water supply issues make diarrhoeal diseases more likely. Measles is rife as people are pushed into smaller habitable spaces. Low levels of vitamin A can also cause malabsorption and immune problems, meaning even less vitamin A can be absorbed and making the individual more susceptible to infections – including measles.

Vitamin D deficiency

Vitamin D deficiency occurs due to poor dietary intake, poor absorption, or inadequate sunlight exposure. Climate change is likely to affect dietary intake of all vitamins and minerals but might also reduce sunlight exposure. It doesn't take much of a temperature rise for people to seek shade or to move indoors. The higher temperatures we are seeing in our warmer world are more likely to cause people to protect themselves from the sun rather than bathe in it. In many parts of the world where vitamin D deficiency is highest, ambient temperatures are too high in the summer to remain in direct sunlight for long.

Low vitamin D levels have various effects on the body. Ocularly, deficiency increases the risk of cataracts. It has also been shown to cause dry eye by altering the tear osmolarity [107]. Studies have indicated that diabetic retinopathy is more severe in the presence of vitamin D deficiency [107, 108]. Deficiency has also been shown to be associated with thinning of the nerve fibre layer of the macula [107, 108].

Vitamin E deficiency

Vitamin E deficiency occurs either through poor nutrition (especially seeds, vegetables and oils) or through malabsorption. It is relatively rare in adults but more common in children who have lower stores. Vitamin E is needed for the correct functioning of the rod cells of the retina. Deficiency presents as night blindness and as with vitamin A will progress if untreated [107]. There is also some evidence that vitamin E deficiency can have an adverse effect upon ARMD. At the other end of the age scale, deficiency has been implicated in retinopathy of prematurity [107].

Vitamin B12 deficiency

Vitamin B12 deficiency is again caused by either poor intake (eggs, fish and milk) or malabsorption. Mild deficiency is actually quite common, with about 6% of the population of the US and UK deficient, but with higher rates in lower-income countries [107]. Mild deficiency is generally sub-clinical, but more severe deficiency interferes with succinyl-CoA synthesis and this can result in an optic neuropathy which may not be reversible. Vitamin B12 has also been shown to cause dry eye disease, which can be severe [109].

Zinc deficiency

Zinc is an essential micronutrient for humans and has vital roles in cell structure and metabolism [110]. Once again, deficiency is caused by poor intake (meat, fish, vegetables) and/or poor absorption. Zinc deficiency is common globally, especially in developing countries and has been designated by the WHO as a major disease [111].

Zinc is essential for the correct functioning of vitamin A so zinc deficiency will cause night blindness. Zinc interacts with taurine and vitamin A in the retina, modifying plasma membranes of the photoreceptors, which regulates the light-rhodopsin reaction within the photoreceptors [112].

Deficiency has also been shown to cause dry eye. It has also been implicated in the pathogenesis of ARMD [112].

Inactivity, obesity, and diabetes

Climate change has effects beyond the most obvious, but this doesn't make them any less of a risk to our health. Physical inactivity is one of the leading risk factors for diseases and death worldwide. It increases the risk of cancer, heart disease, stroke and diabetes by 20–30%. It is estimated that four to five million deaths per year could be averted if the global population was more active [113].

Greater ambient temperatures that *stay within* the temperate parameters (of that area/region) is associated with *increased* physical activity. However, when the ambient temperature *exceeds* the normal parameters it has been found to *reduce* physical activity [114]. Whilst moderate heat might get us to go outside more and be more active, it doesn't need to get much hotter for our hypothalamus to tell us to get out of the heat. In our modern world this usually means either going into our houses or into our cars [114]. These findings are not surprising and indeed are likely to be self-protective in the short term – physical activity raises body temperature itself so high ambient temperatures are to be avoided.

Climate change and obesity are linked. Obesity has been found to be associated with 20% greater emissions of greenhouse gases and climate change is more likely to lead to obesity [115]. The latter relationship is multifactorial. As higher ambient temperature reduces physical activity it therefore reduces calorie expenditure. As climate change effects agriculture, prices of fresh food go up and this may encourage more people to buy ultra-processed food. Ultra-processed food contains far more cheap ingredients such as salt and sugar and fats, and these are also far more obesogenic than unprocessed food [115].

Obesity and inactivity have been linked to a whole host of health problems and eye diseases are no exception. Evidence exists for increased risks of cataract, glaucoma and ARMD [116]. The strongest evidence is the well-established link between obesity and Type 2 Diabetes. 90% of adults with Type 2 Diabetes are overweight or obese [117].

Diabetes can have profound effects on our eyes and vision. Few ocular structures can be unaffected by the complications of diabetes. We will look at the numbers and economics of this in the next chapter, but it represents a huge burden on healthcare. Visually significant cataract is common – with around 20% of all cataract operations being performed on diabetic patients [118]. Diabetic retinopathy is a potentially blinding disease unless treated. Maculopathy affects central vision whilst proliferative diseases can affect the whole eye and results in secondary glaucoma.

The muscles that control eye movement can be affected by diabetes as can the blood supply to the optic nerve. Corneal problems can occur if the corneal nerves are damaged and the corneal sensation reduced. The significant increase in cardiovascular disease in diabetics increases the risk of strokes and ischaemic events, which can affect the vision centres in the brain. Diabetics are more prone to infections and this includes of the eye and its surrounding structures.

Changes in animal behaviour

Climate change inevitably means migration for both plants and animals in an effort to regain their previous environment. As this is invariably to cooler areas this means either heading further north or heading for higher altitude. One study estimated that for a range of species this was around 11 metres per decade to higher elevations and around 17 kilometres per decade further north [119].

Larger animal migration can lead to the added pressure of shrinking space and food and water supply and will increasingly bring humans into contact with these animals. Whilst humans will no doubt win this battle, there will be casualties along the way. Ocular injures from animals are already surprisingly common worldwide, with the type of animal depending upon the region. Domestically, dogs and cats cause the most common ocular injuries. Cattle in India are responsible for a significant number whilst camel injuries are most common in the Middle East. As we encroach upon their habitats more animals will present this threat – reptile and bear injuries are well-described [120]. Ocular injuries from animals can be very serious. If surgery is required it can be complex due to the irregular nature of the injury and the risk of infection is high.

A further risk from animals comes from flooding. Flooding forces animals out of their normal habitat, especially those that live underground. Thus, snakes and spiders are forced en masse to seek out dry land. Spiders and snakes and caterpillars can cause significant ocular injuries. Direct venom sprayed into the eye can cause serious damage to the cornea and ocular surface. Venom injected into the blood can cause severe retinal and optic nerve damage [121].

Migration, population changes, and war

Climate change pressures not only force non-human animals to migrate but are increasingly creating the same pressures in humans. As the heat rises, crops and livestock fail and severe weather events worsen, those who can begin to look for safer environments. For many this may involve only a few miles trek to be nearer a river or sea or to a higher altitude. For others this may mean hundreds of miles of dangerous treks. Climate change is estimated by the World Bank to force up to 216 million people across six world regions (sub-Saharan Africa, South Asia, Latin America, East Asia, and the Pacific, North Africa, Eastern Europe and Central Asia) to leave their home areas by 2050 [122].

The act of migration in itself produces health hazards including ocular injuries, dehydration, and poor nutrition. Migrants may bring infective diseases into a population that has little immunity. Conversely, migrants are exposed to infective diseases they may not have previously encountered. Inevitably, with more people living in smaller areas infectious diseases are far easier to pass on. All the ocular infections mentioned in the previous section can be exacerbated by this.

Large numbers of migrants can put pressures upon local healthcare systems that they may not be equipped to deal with. One study of migrants found that vision impairment was claimed by 16.6% of subjects. This was made up of refractive errors (11%), strabismus (6%), red eye (6%), cataract (5.3%) and ocular hypertension (1%). Retinal pathology was found in 5% of migrants. Nearly 10% of the migrants screened had never previously had an eye test [123].

Migration also can lead to conflict when large numbers of people try to move to where others are already settled. Differences might be cultural or simply cause exhaustion of natural resources. Governments might use these migrations as an excuse to demonise a foreign culture. Countries might fight over water sources or fishing grounds. Climate change will be a potent cause of future inter human conflict and there is no human activity that can cause more eye trauma and blindness than war.

A study by McMaster and Clare found an increasing trend of ocular injury in modern conflict due to weapons with higher explosive and fragmentation power. As well as increasing numbers of ocular injuries this means they are more severe. In modern conflict zones, up to 15.8% of all medical evacuations have sustained eye injuries. In 2019, following a vehicle-borne improvised explosive device detonation in Mogadishu, Somalia ocular injuries were found in 24.6% of survivors [124]. Ocular blast injuries are difficult to deal with and have a high risk of blindness and the numbers who are affected by this (especially civilian) are increasing [125].

References

1. Parker ER. The influence of climate change on skin cancer incidence – A review of the evidence. *International Journal of Women's Dermatology* 2020 Jul 17;7(1):17–27. https://doi.org/10.1016/j.ijwd.2020.07.003.
2. Cancer.net 2023. Eyelid Cancer: Statistics. https://www.cancer.net/cancer-types/eyelid-cancer/statistics.
3. Xia N, Chen L, Chen H, Luo X et al. Influence of atmospheric relative humidity on ultraviolet flux and aerosol direct radiative forcing: Observation and simulation. *Asia-Pacific Journal of Atmospheric Sciences* 2016;52:341–352. https://doi.org/10.1007/s13143-016-0003-2.6,87.
4. Echevarría-Lucas L, Senciales-González JM, Medialdea-Hurtado ME et al. Impact of climate change on eye diseases and associated economical costs. *International Journal of Environmental Research and Public Health* 2021 Jul 5;18(13):7197. https://doi.org/10.3390/ijerph18137197.
5. The Skin Cancer Foundation 2024. *The Heat is On.* https://www.skincancer.org/blog/the-heat-is-on/#:~:text=Since%20the%201940s%2C%20animal%20and,UV%20on%20the%20cellular%20level.
6. Cancer Research UK. Risk and Causes of Skin Cancer. https://www.cancerresearchuk.org/about-cancer/skin-cancer/risks-causes.
7. Bocheva G, Slominski RM, Slominski AT. Environmental air pollutants affecting skin functions with systemic implications. *International Journal of Molecular Sciences* 2023 Jun 22;24(13):10502. https://doi.org/10.3390/ijms241310502.
8. Adamski WZ, Maciejewski J, Adamska K, Marszałek A, Rospond-Kubiak I. The prevalence of various eyelid skin lesions in a single-centre observation study. *Postępy Dermatologii i Alergologii* 2021 Oct;38(5):804–807. https://doi.org/10.5114/ada.2020.95652.
9. DermNet 2014. Eyelid Contact Dermatitis. https://dermnetnz.org/topics/eyelid-contact-dermatitis.
10. Fathy R, Rosenbach M. Climate change and inpatient dermatology. *Current Dermatology Reports* 2020;9:201–209. https://www.ncbi.nlm.nih.gov/pmc/articles/PMC7442546/.
11. Ivanov IV, Mappes T, Schaupp P et al. Ultraviolet radiation oxidative stress affects eye health. *Journal of Biophotonics* 2018 Jul;11(7):e201700377. https://doi.org/10.1002/jbio.201700377.
12. Ziska LH. Climate, carbon dioxide, and plant-based aero-allergens: A deeper botanical perspective. *Frontiers in Allergy* 2021 Aug 20;2:714724. https://doi.org/10.3389/falgy.2021.714724.
13. Nam S, Shin MY, Han JY et al. Correlation between air pollution and prevalence of conjunctivitis in South Korea using analysis of public big data. *Scientific Report* 2022;12:10091. https://doi.org/10.1038/s41598-022-13344-5.
14. Aik J, Chua R, Jamali N et al. The burden of acute conjunctivitis attributable to ambient particulate matter pollution in Singapore and its exacerbation during

South-East Asian haze episodes. *Science of The Total Environment* 2020:740. https://doi.org/10.1016/j.scitotenv.2020.140129.

15. Cancer.net 2023. Eye Melanoma: Statistics. https://www.cancer.net/cancer-types/eye-melanoma/statistics#:~:text=However%2C%20the%20number%20of%20people,the%20United%20States%20in%202023.
16. The University of Manchester 2021. Ultraviolet Radiation Causes Rare Type of Eye Cancer. https://www.manchester.ac.uk/discover/news/ultraviolet-radiation-causes-rare-type-of-eye-cancer/.
17. Liu X, Qiu S, Liu Z et al. Effects of floods on the incidence of acute hemorrhagic conjunctivitis in Mengshan, China, from 2005 to 2012. *American Journal of Tropical Medicine and Hygiene* 2020 Jun;102(6):1263–1268. https://doi.org/10.4269/ajtmh.19-0164.
18. Stapleton F, Alves M, Bunya VY et al. TFOS DEWS II epidemiology report. *Ocular Surface* 2017 Jul;15(3):334–365. https://doi.org/10.1016/j.jtos.2017.05.003.
19. Alves M, Novaes P, Morraye M et al. Is dry eye an environmental disease? *Arquivos Brasileiros de Oftalmologia* 2014;77(3). https://doi.org/10.5935/0004-2749.20140050.
20. Zhong JY, Lee YC, Hsieh CJ et al. Association between dry eye disease, air pollution and weather changes in Taiwan. *International Journal of Environmental Research and Public Health* 2018 Oct 16;15(10):2269. https://doi.org/10.3390/ijerph15102269.
21. Patel S, Mittal R, Kumar N et al. The environment and dry eye-manifestations, mechanisms, and more. *Frontiers in Toxicology* 2023 Aug 23;5:1173683. https://doi.org/10.3389/ftox.2023.1173683.
22. García-Marqués JV, Talens-Estarelles C, García-Lázaro S et al. Systemic, environmental and lifestyle risk factors for dry eye disease in a mediterranean Caucasian population. *Contact Lens and Anterior Eye* 2022 Oct;45(5):101539. https://doi.org/10.1016/j.clae.2021.101539.
23. Mo Z, Fu Q, Lyu D et al. Impacts of air pollution on dry eye disease among residents in Hangzhou, China: A case-crossover study. *Environmental Pollution* 2019 Mar;246:183–189. https://doi.org/10.1016/j.envpol.2018.11.109.
24. Yu D, Deng Q, Wang J et al. Air pollutants are associated with dry eye disease in urban ophthalmic outpatients: A prevalence study in China. *Journal of Translational Medicine* 2019 Feb 15;17(1):46. https://doi.org/10.1186/s12967-019-1794-6.
25. Mandell JT, Idarraga M, Kumar N et al. Impact of air pollution and weather on dry eye. *Journal of Clinical Medicine* 2020 Nov 20;9(11):3740. https://doi.org/10.3390/jcm9113740.
26. Zhong JY, Lee YC, Hsieh CJ et al. Association between dry eye disease, air pollution and weather changes in Taiwan. *International Journal of Environmental Research and Public Health* 2018 Oct 16;15(10):2269. https://doi.org/10.3390/ijerph15102269.
27. Hwang SH, Choi YH, Paik HJ et al. Potential importance of ozone in the association between outdoor air pollution and dry eye disease in South Korea. *JAMA Ophthalmology* 2016 May 1;134(5):503–510. https://doi.org/10.1001/jamaophthalmol.2016.0139.
28. American Academy of Ophthalmology: EyeWiki. Ocular Manifestations of Climate Change. https://eyewiki.org/Ocular_Manifestations_of_Climate_Change.

29. Izadi M, Jonaidi-Jafari N, Pourazizi M et al. Photokeratitis induced by ultraviolet radiation in travelers: A major health problem. *Journal of Postgraduate Medicine* 2018 Jan-Mar;64(1):40–46. https://doi.org/10.4103/jpgm.JPGM_52_17.
30. Delic NC, Lyons JG, Di Girolamo N et al. Damaging effects of ultraviolet radiation on the cornea. *Photochemistry and Photobiology* 2017;93(4):920–929. https://doi.org/10.1111/php.12686.
31. Rezvan F, Khabazkhoob M, Hooshmand E et al. Prevalence and risk factors of Pterygium: A systematic review and meta-analysis. *Survey of Ophthalmology* 2018 Sep–Oct;63(5):719–735. https://doi.org/10.1016/j.survophthal.2018.03.001.
32. Durkin SR, Abhary S, Newland HS et al. The prevalence, severity and risk factors for Pterygium in central Myanmar: The Meiktila Eye Study. *British Journal of Ophthalmology* 2008;92:25–29. https://doi.org/10.1136/bjo.2007.119842.
33. StatPearls Publishing 2024. Ocular Surface Squamous Neoplasia. https://www.ncbi.nlm.nih.gov/books/NBK573082/#:~:text=To%20conclude%2C%20ocular%20surface%20squamous,clinical%20suspicion%20to%20diagnose%20it.
34. Yeoh CHY, Lee JJR, Lim BXH et al. The management of Ocular Surface Squamous Neoplasia (OSSN). *International Journal of Molecular Sciences* 2022 Dec 31;24(1):713. https://doi.org/10.3390/ijms24010713.
35. The International Commission on Non-Ionizing Radiation Protection. Guidelines on limits of exposure to ultraviolet radiation of wavelengths between 180 nm AND 400 nm (incoherent optical radiation). *Health Physics* 2004;87(2):171–186.
36. West S. Ocular ultraviolet B exposure and lens opacities: A review. *Journal of Epidemiology* 1999 Dec;9(6 Suppl):S97–101. https://doi.org/10.2188/jea.9.6sup_97.
37. Lv X, Gao X, Hu K et al. Associations of humidity and temperature with cataracts among older adults in China. *Frontiers in Public Health* 2022 Mar 31;10:872030. https://doi.org/10.3389/fpubh.2022.872030.
38. Sharon N, Bar-Joseph P, Bormusov E et al. Simulation of heat exposure and damage to the eye lens in a neighborhood bakery. *Experimental Eye Research* 2008;87(1):49–55. https://doi.org/10.1016/j.exer.2008.04.007.
39. Minassian DC, Mehra V, Verrey JD. Dehydrational crises: Ā major risk factor in blinding cataract. *British Journal of Ophthalmology* 1989 Feb;73(2):100–115. https://doi.org/10.1136/bjo.73.2.100.
40. Xiaojie L, Jiying X, Jiahong X et al. Long-term exposure to ambient $PM_{2.5}$ and age-related cataracts among Chinese middle-aged and older adults: Evidence from two National Cohort Studies. *Environmental Science & Technology* 2023;57(32):11792–11802. https://doi.org/10.1021/acs.est.3c02646.
41. Chua S, Khawaja A, Desai P et al. The association of ambient air pollution with cataract surgery in UK biobank participants: Prospective Cohort Study. *Investigative Ophthalmology & Visual Science* 2021;62(15):7. https://doi.org/10.1167/iovs.62.15.7.
42. Wang W, He M, Li Z et al. Epidemiological variations and trends in health burden of glaucoma worldwide. *Acta Ophthalmologica* 2019;97:e349–355. https://doi.org/10.1111/aos.14044.
43. Pasquale LR, Jiwani AZ, Zehavi-Dorin T et al. Solar exposure and residential geographic history in relation to exfoliation syndrome in the United States and Israel. *JAMA Ophthalmology* 2014 Dec;132(12):1439–1445. https://doi.org/10.1001/jamaophthalmol.2014.3326.

44. Louwies T, Panis L, Kicinski M et al. Retinal microvascular responses to short-term changes in particulate air pollution in healthy adults. *Environmental Health Perspectives* 2013;121(9):1011–1016. https://doi.org/10.1289/ehp.120572.
45. Chua SYL, Khawaja AP, Morgan J et al. UK biobank eye and vision consortium. The relationship between ambient atmospheric fine particulate matter (PM2.5) and glaucoma in a large community cohort. *Investigative Ophthalmology & Visual Science* 2019 Nov 1;60(14):4915–4923. https://doi.org/10.1167/iovs.19-28346.
46. Sun HY, Luo CW, Chiang YW et al. Association between $PM_{2.5}$ exposure level and primary open-angle glaucoma in Taiwanese adults: A nested case-control study. *International Journal of Environmental Research and Public Health* 2021 Feb 10;18(4):1714. https://doi.org/10.3390/ijerph18041714.
47. Min K-B, Min J-Y. Association of ambient particulate matter exposure with the incidence of glaucoma in childhood. *American Journal of Ophthalmology* 2020;211:176–182. https://doi.org/10.1016/j.ajo.2019.11.013.
48. Lin H-C, Chen C-S, Keller J et al. Seasonality of retinal detachment incidence and its associations with climate: An 11-year nationwide population-based study. *The Journal of Biological and Medical Rhythm Research* 2011;28(10):942–948. https://doi.org/10.3109/07420528.2011.613324.
49. Prabhu PB, Raju KV. Seasonal variation in the occurrence of rhegmatogenous retinal detachment. *The Asia-Pacific Journal of Ophthalmology (Phila)* 2016 Mar–Apr;5(2):122–126. https://doi.org/10.1097/APO.0000000000000129.
50. Kim DY, Hwang H, Kim JH et al. The association between the frequency of rhegmatogenous retinal detachment and atmospheric temperature. *Journal of Ophthalmology* 2020 Jul 22;2020:2103743. https://doi.org/10.1155/2020/2103743.
51. Chang KH, Hsu PY, Lin CJ et al. Traffic-related air pollutants increase the risk for age-related macular degeneration. *Journal of Investigative Medicine* 2019 Oct;67(7):1076–1081. https://doi.org/10.1136/jim-2019-001007.
52. Grant A, Leung G, Aubin MJ et al. Fine particulate matter and age-related eye disease: The Canadian Longitudinal Study on aging. *Investigative Ophthalmology & Visual Science* 2021 Aug 2;62(10):7. https://doi.org/10.1167/iovs.62.10.7.
53. Millen A, Dighe S, Kordas K et al. Air pollution and chronic eye disease in adults: A scoping review. *Ophthalmic Epidemiology* 2024;31(1):1–10. https://doi.org/10.1080/09286586.2023.2183513.
54. Lin CC, Chiu CC, Lee PY et al. The adverse effects of air pollution on the eye: A review. *International Journal of Environmental Research and Public Health* 2022 Jan 21;19(3):1186. https://doi.org/10.3390/ijerph19031186.
55. Shan A, Chen X, Yang X et al. Association between long-term exposure to fine particulate matter and diabetic retinopathy among diabetic patients: A national cross-sectional study in China. *Environment International* 2021:154106568. https://doi.org/10.1016/j.envint.2021.106568.
56. Chalam K, Khetpal V, Rusovici R et al. A review: Role of ultraviolet radiation in age-related macular degeneration. *Eye & Contact Lens: Science & Clinical Practice* 2011;37(4):225–232. *https://doi.org/10.1097/ICL.0b013e31821fbd3e.*
57. Schick T, Ersoy L, Lechanteur YT et al. History of sunlight exposure is a risk factor for age-related macular degeneration. *Retina* 2016 Apr;36(4):787–790. https://doi.org/10.1097/IAE.0000000000000756.

58. Delcourt C, Cougnard-Grégoire A, Boniol M et al. Lifetime exposure to ambient ultraviolet radiation and the risk for cataract extraction and age-related macular degeneration: The Alienor study. *Investigative Ophthalmology & Visual Science* 2014;55(11):7619–7627. https://doi.org/10.1167/iovs.14-14471.
59. Cheng HC, Pan RH, Yeh HJ et al. Ambient air pollution and the risk of central retinal artery occlusion. *Ophthalmology* 2016;123:2603–2609. https://doi.org/10.1016/j.ophtha.2016.08.046.
60. Tan H, Pan S, Zhong Z et al. Association between fine particulate air pollution and the onset of uveitis in Mainland China. *Ocular Immunology and Inflammation* 2022 Oct–Nov;30(7–8):1810–1815. https://doi.org/10.1080/09273948.2021.1960381.
61. Tan H, Pan S, Zhong Z et al. Association between temperature changes and uveitis onset in mainland China. *British Journal of Ophthalmology* 2022 Jan;106(1):91–96. https://doi.org/10.1136/bjophthalmol-2020-317007.
62. Tucker MA, Shields JA, Hartge P et al. Sunlight exposure as risk factor for intraocular malignant melanoma. *The New England Journal of Medicine* 1985 Sep 26;313(13):789–792. https://doi.org/10.1056/NEJM198509263131305.
63. Stålhammar G, Williams PA, Landelius T. The prognostic implication of latitude in uveal melanoma: A nationwide observational cohort study of all patients born in Sweden between 1947 and 1989. *Discover Oncology* 2022 Oct 31;13(1):116. https://doi.org/10.1007/s12672-022-00584-0.
64. Yu GP, Hu DN, McCormick SA. Latitude and incidence of ocular melanoma. *Photochemistry and Photobiology* 2006 Nov–Dec;82(6):1621–1626. https://doi.org/10.1562/2006-07-17-RA-970.
65. Karimaghaei C, Merkley K, Nazari H. Ophthalmology emergency room admission after Hurricane Harvey. *American Journal of Disaster Medicine* 2022 Fall;14(4):255–261. https://doi.org/10.5055/ajdm.2021.0409.
66. Go JA, Lee M, Alexander NL et al. Eyes of a hurricane: The effect of hurricane Harvey on ophthalmology consultations at Houston's County hospital. *Disaster Medicine and Public Health Preparedness* 2023;17:e13. https://doi.org/10.1017/dmp.2020.470.
67. Jaiswal S, Jalbert I, Schmid K et al. Smoke and the eyes: A review of the harmful effects of wildfire smoke and air pollution on the ocular surface. *Environmental Pollution* 2022:309. https://doi.org/10.1016/j.envpol.2022.119732.
68. *The Urban Institute.* Initial Health Policy Responses to Hurricane Katrina and Possible Next Steps. Stephen Zuckerman and Teresa Coughlin February 2006. https://webarchive.urban.org/UploadedPDF/900929_health_policy.pdf.
69. Altherr FM, Nute AW, Zerihun M et al. Associations between water, sanitation and hygiene (WASH) and trachoma clustering at aggregate spatial scales, Amhara, Ethiopia. *Parasites & Vectors* 2019 Nov 14;12(1):540. https://doi.org/10.1186/s13071-019-3790-3.
70. Hägi M, Schémann J-F, Mauny F et al. Active trachoma among children in Mali: Clustering and environmental risk factors. *PLOS Neglected Tropical Diseases* 2010 Jan 19;4(1):e583. https://doi.org/10.1371/journal.pntd.0000583.
71. Phiri I, Manangazira P, Macleod CK et al. The burden of and risk factors for trachoma in selected districts of Zimbabwe: Results of 16 population-based prevalence surveys. *Ophthalmic Epidemiology* 2018 Dec 28;25(sup1):181–191. https://doi.org/10.1080/09286586.2017.1298823.

72. Schémann J-F, Sacko D, Malvy D et al. Risk factors for trachoma in Mali. *International Journal of Epidemiology* 2002;(31):194–201. https://doi.org/10.1093/ije/31.1.194.
73. Tedijanto C, Aragie S, Tadesse Z et al. Predicting future community-level ocular Chlamydia trachomatis infection prevalence using serological, clinical, molecular, and geospatial data. *PLOS Neglected Tropical Diseases*. 2022 Mar 11;16(3):e0010273. https://doi.org/10.1371/journal.pntd.0010273.
74. World Health Organization. Fact Sheets: Trachoma. 2022. https://www.who.int/news-room/fact-sheets/detail/trachoma.
75. Ramesh A, Kovats S, Haslam D et al. The impact of climatic risk factors on the prevalence, distribution, and severity of acute and chronic trachoma. *PLOS Neglected Tropical Diseases* 2013:7(11):e2513. https://doi.org/10.1371/journal.pntd.0002513.
76. El-Sayed A, Kamel M. Climatic changes and their role in emergence and re-emergence of diseases. *Environmental Science and Pollution Research International* 2020 Jun;27(18):22336–22352. https://doi.org/10.1007/s11356-020-08896-w.
77. World Health Organization. Fact Sheets: Onchocerciasis 2022. https://www.who.int/news-room/fact-sheets/detail/onchocerciasis.
78. *StatPearls*. Onchocerciasis. Michael E. Gyasi, Ogugua N. Okonkwo, Koushik Tripathy 2023. https://www.ncbi.nlm.nih.gov/books/NBK559027/.
79. *Academia.edu*. Climate Change and Onchocerciasis: Insights from an Analysis Public Datasets. Ronaldo Figueiró 2023. https://www.academia.edu/112476385/Climate_change_and_onchocerciasis_insights_from_an_analysis_of_public_datasets?uc-sb-sw=39159908.
80. Park YH, Nam HW. Clinical features and treatment of ocular toxoplasmosis. *Korean Journal of Parasitology* 2013 Aug;51(4):393–399. https://doi.org/10.3347/kjp.2013.51.4.393.
81. Kalogeropoulos D, Sakkas H, Mohammed B et al. Ocular toxoplasmosis: A review of the current diagnostic and therapeutic approaches. *International Ophthalmology* 2022 Jan;42(1):295–321. https://doi.org/10.1007/s10792-021-01994-9.
82. Yan C, Liang LJ, Zheng KY et al. Impact of environmental factors on the emergence, transmission and distribution of Toxoplasma gondii. *Parasites & Vectors* 2016 Mar 10;9:137. https://doi.org/10.1186/s13071-016-1432.
83. Esposito MM, Turku S, Lehrfield L et al. The impact of human activities on zoonotic infection transmissions. *Animals* (Basel) 2023 May 15;13(10):1646. https://doi.org/10.3390/ani13101646.
84. Blaizot R, Nabet C, Laghoe L et al. Outbreak of Amazonian toxoplasmosis: A one health investigation in a remote Amerindian Community. *Frontiers in Cellular and Infection Microbiology* 2020;10:401. https://doi.org/10.3389/fcimb.2020.00401.
85. Walkden A, Fullwood C, Tan SZ et al. Association between season, temperature and causative organism in microbial keratitis in the UK. *Cornea* 2018 Dec;37(12):1555–1560. https://doi.org/10.1097/ICO.0000000000001748.
86. *National Institute for Public Health and the Environment*. Climate Change and Recreational Water Related Infectious Diseases. Report 330400002 2010. A.M De Roda Husman, F.M. Shets. https://www.rivm.nl/bibliotheek/rapporten/330400002.pdf.

87. *Centers For Disease Control and Prevention.* Climate Change and Fungal Disease. 2023. https://www.cdc.gov/fungal/climate.html#:~:text=With%20shifting%20temperatures%2C%20fungi%20may,adapted%20to%20surviving%20in%20humans.&text=Heat%20may%20also%20cause%20other,of%20fungi%20to%20infect%20people.
88. Brown L, Leck AK, Gichangi M et al. The global incidence and diagnosis of fungal keratitis. *The Lancet Infectious Diseases* 2021;21(3):E49–E57. https://doi.org/10.1016/S1473-3099(20)30448-5.
89. Saad-Hussein A, El-Mofty HM, Hassanien MA. Climate change and predicted trend of fungal keratitis in Egypt. *Eastern Mediterranean Health Journal* 2011 Jun;17(6):468–473. PMID: 21796962.
90. Casadevall A, Kontoyiannis DP, Robert V. On the emergence of Candida auris: Climate change, azoles, swamps, and birds. *mBio* 2019;10(4):e01397–19.
91. *World Health Organization COP24 Special Report.* Health and Climate Change. https://iris.who.int/bitstream/handle/10665/276405/9789241514972-eng.pdf?sequence=1.
92. El-Sayed A, Kamel M. Climatic changes and their role in emergence and re-emergence of diseases. *Environmental Science and Pollution Research International* 2020 Jun;27(18):22336–22352. https://doi.org/10.1007/s11356-020-08896-w.
93. Merle H, Donnio A, Jean-Charles A et al. Ocular manifestations of emerging arboviruses. *Journal Français d'Ophtalmologie* 2018;41:e235–e43.
94. Dikid T, Jain SK, Sharma A, Kumar A, Narain JP. Emerging & re-emerging infections in India: An overview. *Indian Journal of Medical Research* 2013;138(1):19–31. PMID: 24056553; PMCID: PMC3767269.
95. *World Health Organization.* Global Report on Neglected Tropical Diseases 2023. https://www.who.int/teams/control-of-neglected-tropical-diseases/global-report-on-neglected-tropical-diseases-2023.
96. *World Health Organization.* Neglected Tropical Diseases. 2024. https://www.who.int/news-room/questions-and-answers/item/neglected-tropical-diseases.
97. MacFadden DR, McGough SF, Fisman D et al. Antibiotic resistance increases with local temperature. *Nature Climate Change* 2018 Jun;8(6):510–514. https://doi.org/10.1038/s41558-018-0161-6.
98. Rowe FJ, Hepworth LR, Howard C et al. High incidence and prevalence of visual problems after acute stroke: An epidemiology study with implications for service delivery. *PLoS One* 2019 Mar 6;14(3):e0213035. https://doi.org/10.1371/journal.pone.0213035.
99. Desai Y, Khraishah H, Alahmad B. Heat and the heart. *Yale Journal of Biology and Medicine* 2023 Jun 30;96(2):197–203. https://doi.org/10.59249/HGAL4894.
100. Thangavel P, Park D, Lee YC. Recent insights into particulate matter ($PM_{2.5}$)-mediated toxicity in humans: An overview. *International Journal of Environmental Research and Public Health* 2022 Jun 19;19(12):7511. https://doi.org/10.3390/ijerph1912751.
101. *World Wildlife Fund.* Sustainable Seafood. https://www.worldwildlife.org/industries/sustainable-seafood.
102. Pereira A, Adekunle RD, Zaman M, Wan MJ. Association between vitamin deficiencies and ophthalmological conditions. *Clinical Ophthalmology* 2023 Jul 19;17:2045–2062. https://doi.org/10.2147/OPTH.S401262.

103. *American Academy of Ophthalmology*. What is Vitamin A Deficiency? Kiersten Boyd 2023. https://www.aao.org/eye-health/diseases/vitamin-deficiency#:~:text=Vitamin%20A%20deficiency%20is%20the,year%20of%20losing%20their%20sight.
104. *World Health Organization*. Vitamin A Deficiency. https://www.who.int/data/nutrition/nlis/info/vitamin-a-deficiency#:~:text=Vitamin%20A%20deficiency%20results%20from,rarely%20seen%20in%20developed%20countries.
105. Huiming Y, Chaomin W, Meng M. Vitamin A for treating measles in children. *Cochrane Database of Systematic Reviews* 2005 Oct 19;2005(4):CD001479. https://doi.org/10.1002/14651858.CD001479.pub3.
106. Gilbert C. The eye signs of vitamin A deficiency. *Community Eye Health* 2013;26(84):66–67. PMID: 24782581; PMCID: PMC3936686. https://www.ncbi.nlm.nih.gov/pmc/articles/PMC9810943.
107. Pereira A, Adekunle RD, Zaman M, Wan MJ. Association between vitamin deficiencies and ophthalmological conditions. *Clinical Ophthalmology* 2023 Jul 19;17:2045–2062. https://doi.org/10.2147/OPTH.S401262.
108. Daldal H, Gokmen Salici A. Ocular findings among patients with Vitamin D deficiency. *Cureus* 2021 May 21;13(5):e15159. https://doi.org/10.7759/cureus.15159.
109. Ozen S, Ozer MA, Akdemir MO. Vitamin B12 deficiency evaluation and treatment in severe dry eye disease with neuropathic ocular pain. *Graefe's Archive for Clinical and Experimental Ophthalmology* 2017 Jun;255(6):1173–1177. https://doi.org/10.1007/s00417-017-3632-y.
110. StatPearls. Zinc Deficiency. Luke Maxfield; Samarth Shukla; Jonathan S. Crane. 2023. https://www.ncbi.nlm.nih.gov/books/NBK493231/.
111. Maywald M, Rink L. Zinc in human health and infectious diseases. *Biomolecules* 2022 Nov 24;12(12):1748. https://doi.org/10.3390/biom1212174.
112. Grahn BH, Paterson PG, Gottschall-Pass KT, Zhang Z. Zinc and the eye. *Journal of the American College of Nutrition* 2001 Apr;20(2 Suppl):106–118. https://doi.org/10.1080/07315724.2001.10719022.
113. World Health Organization. Physical Activity. https://www.who.int/health-topics/physical-activity#tab=tab_2.
114. Lanza K, Gohlke J, Wang S, Sheffield PE, Wilhelmi O. Climate change and physical activity: Ambient temperature and urban trail use in Texas. *International Journal of Biometeorology* 2022 Aug;66(8):1575–1588. https://doi.org/10.1007/s00484-022-02302-5.
115. Koch CA, Sharda P, Patel J, Gubbi S, Bansal R, Bartel MJ. Climate change and obesity. *Hormone and Metabolic Research* 2021 Sep;53(9):575–587. https://doi.org/10.1055/a-1533-286.
116. Cheung N, Wong TY. Obesity and eye diseases. *Survey of Ophthalmology* 2007 Mar–Apr;52(2):180–195. https://doi.org/10.1016/j.survophthal.2006.12.003.
117. Public Health England. Adult Obesity and Type 2 Diabetes. 2014. https://assets.publishing.service.gov.uk/media/5a7f069140f0b6230268d059/Adult_obesity_and_type_2_diabetes_.pdf.
118. Kelkar A, Kelkar J, Mehta H, Amoaku W. Cataract surgery in diabetes mellitus: A systematic review. *Indian Journal of Ophthalmology* 2018 Oct;66(10):1401–1410. https://doi.org/10.4103/ijo.IJO_1158_17.
119. Chen IC, Hill JK, Ohlemüller R et al. Rapid range shifts of species associated with high levels of climate warming. *Science* 2011 Aug 19;333(6045):1024–1026. https://doi.org/10.1126/science.1206432.

120. https://www.researchgate.net/publication/358432199_Animal_induced_ocular_injuries_A_brief_review.
121. Maurya R, Singh V, Narayan S et al. Animal induced ocular injures: A brief review. *International Journal of Ocular Oncology and Oculoplasty* 2021;7(4):335–343.
122. The World Bank. Climate Change Could Force 216 Million People to Migrate Within Their Own Countries by 2050. https://www.worldbank.org/en/news/press-release/2021/09/13/climate-change-could-force-216-million-people-to-migrate-within-their-own-countries-by-2050.
123. Bruscolini A, Visioli G, Marenco M et al. Eye health screening in migrant population: Primary care experience in Lazio (Italy) from the PROTECT project. *Applied Sciences* 2023;*13*:3618. https://doi.org/10.3390/app13063618.
124. McMaster D, Clare G. Incidence of ocular blast injuries in modern conflict. *Eye* (Lond) 2021 Dec;35(12):3451–3452. https://doi.org/10.1038/s41433-020-01359-z.
125. Elbeyli A, Kurtul BE. A series of civilian ocular injuries from the civil war in Syria. *Beyoglu Eye Journal* 2020 Dec 28;5(3):205–208. https://doi.org/10.14744/bej.2020.5866.

4 Climate change and eye care

It is a recurring theme in this book that climate change will affect every corner of our world. It is difficult to envisage any aspect of human health or behaviour that will not eventually be affected by this change. We saw this in the previous chapter which outlined some of the major threats to our eyes and vision. There are few eye conditions that will not be directly or indirectly worsened by a warming world. There may also be new diseases that appear or a resurgence of ancient diseases that we hoped were lost (for example the reappearance of anthrax with the melting of the permafrost).

That these eye diseases can cause misery to the sufferers is one aspect, but they will also put greater and greater pressures upon healthcare resources. Even mature economies are struggling to deal comprehensively with their population's healthcare needs and this pressure can only increase. Many developing countries are unable to provide even basic healthcare to some parts of their population and these are the countries climate change will have its greatest initial effect upon.

This chapter is about the current burden of eye care on healthcare systems and governments and projections of how much this will grow. The economics of healthcare matters as there is a limited amount of money to go round. The financial needs of one speciality or disease are another's loss. Not all eye conditions can be covered so the chapter is focused upon the most prevalent ones and/or those most likely to be impacted by climate change. Thus, these figures will be an underestimate of the entire number of sufferers and the costs of their disease. It is clear we cannot as yet know the full ramifications of climate change on our health. Some conditions we can predict, some we can't and some diseases may appear or reappear unexpectedly. This lack of predictability is one the reasons why climate change needs to be urgently addressed before we will lose control of our futures.

The WHO *World Report on Vision* states that globally at least 2.2 billion people have a visual impairment or are blind. The visually disabling conditions that are the simplest and most effective to fix – refractive errors and cataracts – are calculated to have a funding gap of $24.8 billion US dollars.

DOI: 10.4324/9781003512608-5

This funding gap means how much it is *projected* it would cost to correct all those with refractive errors and cataract, *not* how much is currently spent [1].

The Report states, 'In the coming decades, if the projected increase in older people is not met with increased access to eye care services, there will be a substantial increase in the number of people with vision impairment and blindness'. This means financial investment is needed immediately to clear the current backlog and prepare for the future demand. It requires appropriate planning to ensure this additional investment will strengthen existing health systems. The WHO has estimated that in order to achieve the global health targets set for 2030, low- and middle-income countries will need to invest in an additional 23 million health workers and build more than 415,000 new health facilities. This will require an estimated US$24.8 billion in additional investment if these workforce and infrastructure needs are to be met [2].

Whilst demand for eye care will increase inexorably even without the added burden of climate change, there is no guarantee that this demand will be met, and it may even worsen. A study from the US calculated that from 2020 to 2035, the total *supply* is projected to decrease by 2,650 full-time equivalent (FTE) ophthalmologists (a 12% decline) and total *demand* is projected to increase by 5,150 FTE ophthalmologists (a 24% increase). This is a supply and demand mismatch of 30% of the necessary workforce. The USA has some of the largest number of ophthalmologists globally so this gap is likely to be far greater in other parts of the world – both North and South [3].

We will now look at some of the major eye conditions, how much they currently cost and the future projections of numbers and estimated costs.

4.1 Eye associated cancers

According to the World Cancer Research Fund International there were 324,635 melanoma skin cancers diagnosed in 2020 globally. This represented an age-standardised prevalence of 3.4/100,000 of the population – with the highest numbers in Australia and New Zealand. For non-melanoma skin cancers there were 1,198,073 cases also in 2020. This represented an age-standardised prevalence of 11/100,000 population, again with the greatest number of cases in Australia and New Zealand [4].

From the same data, in 2020 there were 57,043 deaths from melanoma skin cancers with the highest totals in Australia. There were 63,371 non-melanoma skin cancer deaths in the same year, with the highest total deaths in Papua New Guinea.

Eyelid tumours are between 5–10% of all skin cancers [5]. Using the figures above, the total number of all skin cancer cases was 1,522,708, which would mean eyelid cancer numbers in 2020 would be up to 150,000 worldwide. Around 90% of eyelid tumours are BCCs so that would mean around

135,000 cases. 5% are SCCs so that would be around 7,500 cases and melanomas represent around 1% so that would be around 1,500 cases.

There is good evidence that these eyelid tumours are increasing in number – as we would expect from greater UV radiation, air pollution, and greater ambient temperatures. Lin *et al.* found a rapidly increasing incidence of eyelid cancers from 1979 to 1999 in Taiwan [6]. Quigley *et al.* in Ireland also found increasing numbers, especially in women and those under 50 years of age [7]. Similarly, Lee *et al.* found an increasing incidence in women in Singapore [8]. Safi *et al.* found in an Iranian population an increasing trend of eyelid cancers over the previous decade [9]. Jung *et al.* found an increase in numbers between 1993 and 2016 in Korea [10].

As the prevalence of skin cancers has increased so have the associated costs. A study from the US estimated annual costs rose from US$8.0 billion in 2012–2015 to US$8.9 billion in 2016–18 [11]. Another US study calculated that in 2007–2011 nearly 5 million adults were treated for skin cancer annually. The average treatment costs were US$8.1 billion in 2011 compared with US$3.6 billion in 2007. This represented an increase of 126.2%, while the average annual total cost for all other cancers increased by 25.1% [12].

What aspects of this treatment are the main cost drivers? A Swedish study reported that the total cost of skin cancer was estimated to be €142.4 million, of which €79.6 million was spent on health services and €62.8 million was due to loss of productivity. The main cost driver was outpatient care, amounting to 42.2% of the total cost [13].

Another obvious eye-related tumour that could be affected by climate change is uveal melanoma. *Cancer.net* gives the overall number of melanoma cases (from all sites) as 324,635 worldwide [14]. Of all melanoma cases, uveal melanoma (principally choroidal) represents 3–5% of cases. This would mean up to 16,000 people were diagnosed with uveal melanoma in 2020. The 5-year relative survival rate for people with large choroidal melanoma is 47% [15].

Overall uveal melanoma numbers have not been shown to be increasing as yet but there is evidence that some people will be at more risk than others. An Australian study found that this type of melanoma was more common in outdoor workers and that individual sun-exposure measurements suggest that solar radiation was causative in ocular melanoma [16].

Uveal melanoma treatment needs specialist treatment from specifically trained practitioners as well as lifelong follow up. There are few estimates of the costs of this, but one paper from the US does try to quantify it. They found that removal of the eye (enucleation) had the lowest cost at US$22,772 per patient. External proton beam therapy had a cost of US$24,894 whilst plaque therapy (stitching a radioactive plaque on the sclera overlying the tumour) was the most expensive at US$28,862 [17]. If there has been metastatic spread of the melanoma, a relatively new immunotherapy drug, tebentafusp (*Kimmtrak*), was approved by the FDA in January 2022. It is estimated that it costs between US$400,000 and US$500,000 for a year's course of treatment [18].

Finally, although conjunctival melanomas make up less than 1% of melanomas, there is evidence of a marked increase in their incidence. Yu *et al.* found that conjunctival melanoma incidence increased by 5.5% biannually from 1973 to 1999, and suggested that the change in incidence of conjunctival melanoma paralleled that of cutaneous melanoma and had a possible aetiology of exposure to sunlight [19].

4.2 Dry eye disease

Population studies using standardised definitions of dry eye disease have found a wide range of prevalence from 5–50% [20]. This prevalence increases with age, and women are at higher risk than men. Studies have shown significant personal and economic burdens of dry eye on vision, pain, quality of life, and work productivity.

Based on data from the *National Health and Wellness Survey*, 6.8% of the US adult population (approximately 16.4 million people) have been diagnosed with dry eye disease [21]. The prevalence increases with age (2.7% in those 18 to 34 years old versus 18.6% in those over 75 years). This means significant numbers of people who are economically active suffer significant dry eye disease. A 2022 meta-analysis of three studies from the US estimated a prevalence of dry eye of 8.1% [22].

We saw in the previous chapter that climate change will directly impact upon the ocular surface through various mechanisms – increased temperature, allergens, air pollution, wildfires and UV radiation [23]. Projections of future increases in dry eye prevalence are therefore likely to be correct. There are other factors that determine this prevalence – most obviously our (currently) ageing population – but environmental factors are unquestionably significant and likely to be increasingly so [24].

The costs to the individual and the healthcare system from dry eye disease are significant. A Canadian study calculated annual costs of dry eye were US$24,331 Canadian dollars per patient and increased with patient-reported disease severity. Indirect and direct costs increased through mild to moderate to severe dry eye [25]. A US study found that the average annual cost (whoever was paying) of managing a patient with dry eye was US$783 (variation US$757–US$809). When adjusted for the prevalence of the condition nationwide, the overall burden of dry eye disease for the US healthcare system would be US$3.84 billion [26]. These costs are made up of direct costs (outpatient visits, medications) and indirect costs (time of work or unemployment). The indirect costs comprise most of the overall economic burden. For example, in the US, the mean average indirect cost to society was calculated as US$55.4 billion compared with US$3.8 billion for direct costs [26].

Looking at just one aspect – dry eye treatments – it has been calculated that this market was worth US$5,120 million in 2021 and has been projected to rise by 5% annually [27].

4.3 Cataracts

Cataract formation is almost ubiquitous with ageing – one study gave a prevalence of cataracts as 92.6% of the population over 80 [28]. Figures from the 2010 *Global Burden of Disease* reported that worldwide 10.8 million people were blind and 35.1 million were visually impaired due to cataracts. Cataracts were responsible globally for 33.4% of all blindness in 2010. Thus in 2010 one in three blind people were blind due to cataracts, and one in six visually impaired people were visually impaired due to cataracts [29].

It is hard to know the true incidence of cataracts. Minassian and Mehra estimated that for India alone 3.8 million people become blind from cataracts each year [30]. Globally the incidence figure is probably at least 5 million. A figure of 1000 new blind people from cataracts per million population per year is used for planning purposes in developing countries [31].

As we saw in the previous chapter, climate change is likely to have a direct impact upon the development of visually impairing cataracts. From higher ambient heat and UV light to more diabetes, cataracts will become even more common and probably at younger ages. It has been estimated that, just from UV radiation, there will be 200,000 additional cataract cases by 2050. This is more than would be expected even with an ageing population [32].

The cataract surgery rate is the number of cataract operations per year per million population. This ranges from 5,500 in North America to 4,000 in Western Europe to 3,100 in India to 1,000 in the rest of Asia to 300 in Africa. Needless to say, the number of people who have sight-threatening cataracts in each of these regions does not mirror this surgical rate [31]. The paper that this data is taken from puts these figures in perspective:

> In order to reduce the cataract backlog it is necessary to have a cataract surgical rate which is at least as great as the incidence of 'operable' cataract, where 'operable' depends upon the indication for surgery. In India and other countries of South East Asia, in order to deal with cataract causing an acuity less than 6/60, it is necessary to do at least 3000 operations per million population per year and perhaps more. In Africa, and other parts of the world where there is a lower percentage of elderly people in the population, a realistic target for the next 5–10 years is around 2000 operations/million population /year.

At the present time, in many parts of the world, the cataract surgical rate is nowhere near the level needed to prevent significant visual disability. Lower-income regions have the highest disparity between those who need cataract surgery and the availability. These same regions, as we have seen, are also those at greatest risk of the adverse effects of climate change.

The WHO report that to deal with all those globally with cataracts at the current time would cost an estimated US$8.8 billion [1]. In the US alone around US$11 billion is spent on cataract surgery per year [33]. A Canadian study combined medium population growth projections with 2006–2007 cataract surgical rates and estimated a 128% growth in the number of cataract operations by 2036. The proportion of cataract operations provided for older patients was projected to increase significantly, with the number of operations for patients aged 85 years and older more than tripling by 2036. As the authors state, these numbers are likely to be similar all over the developed world. They also note that this does not include a number of other possible pressures – including environmental – which might increase this further [34].

Conversely, the cost of cataract surgery in developing economies may not be so great. As cataract surgery is available to fewer people in these countries overall costs are lower. What this actually means of course is that the gap between those who need cataract surgery and those who get it is immense, unlike in the US or Europe for example. The costs of each case, however, are usually much lower than in those countries [35].

4.4 Age-related macular degeneration

Age-related macular degeneration (ARMD or AMD) is the leading cause of irreversible blindness of those over 65 years of age. Worldwide the number of people with ARMD was around 196 million in 2020. This is estimated to increase to 288 million in 2040 [36]. Another report calculated the worldwide cost of visual impairment due to age-related macular degeneration alone at US$343 billion including US$255 billion in direct healthcare costs [37].

90% of ARMD is of the dry type and at present there is no way to treat this. Specific vitamin supplements may slow this type of degeneration but only time will tell if this has a significant impact upon this form of the disease.

The remaining 10% of ARMD is the so-called wet type. This is called wet as abnormal blood vessels form at the centre of the retina and then bleed. The treatment of wet ARMD has been revolutionised in the last few years by the development of anti-VEGF treatments. These involve regular injections into the eye which block the development of the abnormal blood vessels. Though efficacious, these treatments come at a cost. Studies in the US have shown a substantial and increasing cost of ARMD following the advent of these treatments.

Qualls *at al.* found that the mean cost per ARMD case in the year after diagnosis was US$12,422, which was US$4,884 higher than the year before diagnosis. This increase was attributable primarily to the introduction of intravitreal injections of the vascular endothelial growth factor antagonists. Intravitreal injections averaged US$203 for patients diagnosed in 2004 and

US$2,749 for patients diagnosed in 2008 [38]. In another study, total all-cause costs among anti-VEGF treatment-naïve and previously treated patients with wet AMD over 12 months were, on average, US$21,125 and US$23,096 respectively [39]. These costs are mainly the cost of injection (drug costs and facility costs) and the cost of repeated follow up and investigation. Whilst before the advent of these injections patients were often discharged, they will now have many appointments per year.

As we saw in the previous chapter, ARMD is likely to be negatively affected by elements of climate change. Thus in the future we will see increasing blindness, disability, and the personal and healthcare costs associated with dry ARMD and vastly increasing costs from the treatment of wet ARMD. There may be some evidence from Spain that this is already occurring, with the number of cases of ARMD rising faster than had been predicted [40].

4.5 Infectious diseases and the eye

We have already seen that our warming world will create an increasingly ideal environment for infectious diseases. Everything points to an increase – from the direct effects of bacterial and fungal infections to the indirect effects of flooding, malnutrition, and pollution. It would be hard to think of an infectious disease anywhere in the body that will not threaten us in the future. Eye infections from the lid to conjunctivitis to intraocular infections are all likely to increase.

As we saw in the previous chapter, two of the most common but sight-threatening infectious diseases in the world are trachoma and onchocerciasis and these well illustrate the risks that climate change brings. They also give some hope (see Chapter 7) as treatment and eradication strategies have already been successful. As we saw in Chapter 3, fungal infections have been with us for eons but as the climate changes they are likely to increasingly threaten us.

Trachoma

Trachoma is a leading cause of blindness in 42 countries, with 125 million people living in trachoma endemic areas. The WHO estimates that trachoma is responsible for visual impairment or blindness in 1.9 million people, and this represents about 1.4% of all blindness worldwide [41].

Trachoma is hyperendemic in many of the poorest and most rural areas of Africa, Central and South America, Asia, Australia, and the Middle East. In areas where trachoma is endemic, active infection is found in up to 90% of preschool-aged children [42]. The economic burden of trachoma upon affected individuals and communities is enormous. This is due to productivity loss from blindness and visual impairment and is estimated at US$2.9–5.3 billion annually, increasing to US$8 billion when trichiasis is included.

Contrast these figures with the cost of treatment of trachoma. One study from India found that the direct costs for treatment of corneal opacity was only US$116 per person. A study from the Gambia found that trachoma surgery costs were around US$9 per person [43]. In 2021, according to the WHO, 69,266 people received surgical treatment for advanced stages of the disease, and 64.6 million people were treated with antibiotics. Global antibiotic coverage in 2021 was 44% [41].

We have already seen that the ideal conditions for trachoma to increase and spread are consistent with climate change. Increased temperatures, water supply problems, communities living closer together, and malnutrition all contribute to the risk of infections. However, as we will see in a later chapter some campaigns have been successful in controlling and even eradicating trachoma in some areas. The balance of new cases versus continuing efforts to control them will mean that an increase in the disease is not inevitable. It is very important to remember though that any eradication program will fail without an acceptable level of sanitation, water disposal and water quality [44].

One final warning. The vector fly that carries trachoma – *Musca sorbens* – has been shown to be sensitive to environmental temperature and rainfall. So as global temperatures rise and rainfall patterns change the countries where trachoma occurs will change [45]. As with so many aspects of climate change, healthcare providers in these new regions will have to be ready to respond.

Onchocerciasis

Onchocerciasis is the second most common infectious cause of blindness in the world. The *Global Burden of Disease Study* estimated in 2017 that at least 220 million people required treatment for onchocerciasis. 14.6 million of the infected people already had skin disease and 1.15 million had visual loss. More than 99% of infected people live in just 31 African countries. The disease also exists in two countries in Latin America (the Yanomami area in Brazil and Venezuela) and Yemen [46].

The proportion of people affected by the diseases in endemic areas is high, with about 37 million people in tropical Africa and 140,000 in Latin America, giving an incidence of approximately 40,000 cases per year. Onchocerciasis causes 46,000 new cases of blindness annually, resulting in 270,000 individuals being blinded and an additional 500,000 developing visual impairment [47].

As with trachoma, environmental factors are a significant risk factor for the disease, especially water supply, hygiene, and malnutrition. There is already evidence that increased average temperature increases the risk of onchocerciasis. Figueiro and Docile looked at data from South America and

Africa and concluded that climate change was already increasing the number of cases [48].

Effective treatment for onchocerciasis is available. So although climate change is very likely to increase the number of cases, this could be ameliorated by preventative measures and more targeted treatment. Control of onchocerciasis has been calculated to cost around US$47 million but to eliminate and eradicate would cost over US$200 million. However, eradication would result in financial savings for healthcare systems and individuals and this has been estimated to be around US$12.5 billion up to 2045 [49]. As climate change will increase the overall number of cases these costs are likely to be even higher.

Fungal keratitis

Fungal keratitis – infection of the cornea – is a challenging condition to treat. One study estimated that the annual incidence was over one million cases worldwide. The highest rates were found to be in Asia and Africa. Of these, over 10% of patients had to have the eye removed, which represents an annual loss of over 80,000 eyes [50].

Fungal keratitis is also associated with significant monetary costs per infection. Treatment often requires multiple drugs, significant clinic costs and has a high failure rate [51]. A Thai study found fungal infections entailed the highest total cost of treatment of any corneal infection [52]. One commercial analysis found that the global fungal eye infection market could reach USD 879.45 million by 2030 [53].

As we saw in the previous chapter, climate change is increasingly creating a world more suitable to pathogenic fungi. As a consequence, fungal infections worldwide seem to be getting more common [54, 55]. Ocular fungal infections are no different and show an increasing incidence and geographical spread.

Acanthamoeba

Acanthamoeba keratitis also seems to be becoming more common. Zhang *et al.* noted an increasing number of cases per year [56]. The current incidence has been reported as between 1 and 33 cases per million contact lens wearers in developed countries [57]. In developing countries cases are more likely to occur after corneal trauma in the presence of acanthamoeba, e.g. when swimming or washing in rivers. As the environment for acanthamoeba to flourish and enter waterways becomes more conducive it is likely there will be more cases of this difficult to treat infection.

4.6 Vitamin A deficiency

One of the more insidious but no less harmful effects of climate change will be that of nutritional deficiency. Floods, droughts, heat, wildfires, and predator threats will combine to alter harvest seasons, kill plants, and reduce the nutritional levels of arable crops. This will have the obvious result of famine and the less obvious results of vitamin and mineral deficiencies – initially in children who have lower stores and then later in adults.

Vitamin A deficiency is the leading cause of preventable blindness in children worldwide. It is estimated that 4 million children under five years of age are affected by xerophthalmia, most of whom are in South Asia and sub-Saharan Africa [58]. An estimated 250,000 to 500,000 children become blind every year because of vitamin A deficiency. Half of these children die within a year of losing their sight [59].

In 2004, approximately 190 million children received at least one high-dose vitamin A supplement, representing global coverage of 68 per cent. Supplementation programmes in the highest-risk populations reached approximately 75 per cent of targeted children with at least one capsule. However, two annual doses are necessary to achieve full protection and millions of children are still not fully protected. It is estimated that only 26 of the 103 priority countries attained effective coverage levels in 2004 [60].

UNICEF recommends two doses of the supplement per year up to the age of 6. The supplements themselves are very cheap, and even if transport and labour costs are added, are usually less than US$1 per capsule [61]. Although in the past twenty years there have been significant increases in global vitamin A supplementation, coverage now seems to have stagnated. The hardest to reach children (who are often the most at need) remain so and this is especially when the specific treatment campaigns come to an end [62].

Once more we can see a blinding disease that is highly likely to be adversely affected by climate change. It is also a disease with a cheap and highly effective treatment. However, some populations are not consistently receiving this treatment and these are the populations that climate change will initially hit hardest. Great strides have been made in reducing Vitamin A deficiency as we will see in Chapter 7 but with increasing numbers of cases and stretched healthcare will this progress be maintained?

4.7 Diabetic eye disease

Physical inactivity is a risk factor in itself for death and illness. The WHO report that insufficient physical activity is the 4th leading risk factor for mortality. Approximately 3.2 million deaths and 32.1 million DALYs (DALYS represent years of sub-optimal health) are attributable to low physical activity. This represents a 20–30% increased risk of all-cause mortality for those who are sedentary when compared to those who engage in at least 30 minutes of moderate intensity physical activity most days of the week. In 2008 it was estimated that 31.3% of persons had less than 150 minutes of moderate activity per week [63].

The WHO also report that in 2022 1 in 8 people in the world were obese, which represented a doubling since 1990. 2.5 billion adults in 2022 were classified as overweight, of which 890 million were obese. Over 390 million children and adolescents aged 5–19 years were overweight in 2022, including 160 million who were living with obesity This means that 43% of adults aged 18 years and over were overweight [64].

Physical inactivity and obesity are both independent risk factors for Type 2 Diabetes [65]. There is evidence that they act together to increase the risk of diabetes beyond what would be expected by simply adding the risks. Thus lack of physical activity and obesity seem to have a synergistic effect upon metabolism that makes diabetes more likely [65].

It is unsurprisingly therefore that the global diabetes prevalence is relentlessly increasing. The International Diabetes Federation estimated in 2012 that in 20–79 year olds it was 10.5% (536.6 million people). This is projected to rise to 12.2% (783.2 million) in 2045. This prevalence was found to be higher in urban (12.1%) than rural (8.3%) areas and in high-income (11.1%) compared to low-income countries (5.5%). The greatest projected increase up to 2045 is expected to occur in middle-income countries compared to high and low-income countries. 90% of this is Type 2 Diabetes [66].

All of this comes at a personal and healthcare cost. Global diabetes-related health expenditures were estimated at 966 billion USD in 2021 and are projected to reach 1,054 billion USD by 2045 [67]. Just in the UK it has been calculated that diabetes costs the NHS over £1.5m an hour or 10% of the NHS budget for England and Wales. This equates to over £25,000 being spent on diabetes every minute. In total, an estimated £14 billion pounds is spent per year on treating diabetes and its complications, with the cost of treating the complications representing the much higher cost [68]. It has been calculated in the US that the average economic cost per person was US$13,240 for diagnosed diabetes, US$4,250 for undiagnosed diabetes and US$500 for prediabetes [69].

Over 90 million diabetics have diabetic eye disease. This represents around 35% of diabetics worldwide [70]. The lowest prevalence is in Europe at 20.6% and South-East Asia at 12.5% and highest in Africa at 33.8%, Middle East and North Africa 33.8%, and the Western Pacific region at 36.2% [71]. Diabetic retinopathy is projected to affect 16 million people by 2050 in the US alone, which is a tripling of the current level [72].

A German study found that the total cost of diabetic retinopathy from a societal perspective was €3.51 billion in 2002. The costs increased as the retinopathy progressed, being highest in patients with proliferative retinopathy and lowest in patients with mild, non-proliferative diabetic retinopathy [73]. In the US, the total direct medical costs of diabetic retinopathy in adults have been estimated at US$493 million per year and the average treatment cost per patient at US$629 per year. This is made up of direct medical costs and loss of economic productivity [74]. These costs are all projected to increase as the numbers with Type 2 Diabetes increase. A meta-analysis suggested this cost increase would disproportionately affect countries in the Middle East and North Africa and the Western Pacific [75].

4.8 Glaucoma

Glaucoma (of all types) already has a significant number of sufferers. Tham *et al.* estimated that in 2013 the number of people (aged 40–80 years) with glaucoma worldwide was nearly 65 million [76]. This increased to 76.0 million in 2020 and they projected 111.8 million in 2040. Part of this is related to ageing populations, but if the pressures of climate change do worsen glaucoma, these numbers will be greater [76]. Wang *et al.* also found that the health burden from glaucoma had increased over the last 25 years and that this increase was related to lower socio-economic level, older age, higher ambient ultraviolet radiation and higher levels of air pollution [77].

4.9 Ocular trauma

Ocular trauma is already a significant cause of visual loss worldwide. According to the World Health Organization (WHO) Program for the Prevention of Blindness, it is estimated that 55 million people sustain ocular trauma/disability annually. This results in 19 million with unilateral blindness and 2.3 million with bilateral reduced visual acuity [78]. 24 million people in the US have suffered ocular injuries, of which 1.5 million are visually impaired and 147,000 are blind [79]. The incidence of blindness caused by ocular trauma has been estimated at 9/100,000 individuals in developed countries and 75/100,000 individuals in developing countries. Globe injuries are projected to occur in 3.5/100,000 people worldwide, resulting in approximately 203,000 new cases each year [80]. Li *et al.* have looked at worldwide data and found that the absolute number of eye injuries has substantially increased over the last 30 years [81].

We cannot definitely ascribe this increase to climate change but the greater energy in our weather systems from climate change is highly likely to have an effect on trauma numbers. As we saw previously, wind events, floods and wildfires all have the potential to wreak havoc on one of our most sensitive and important but exposed organs.

4.10 Climate change and the provision of eyecare services

Climate change will not only have an impact on the prevalence and incidence and consequences of eye diseases but will also affect the way it is delivered. Hospitals, primary care centres, eye camps, medical and nursing staff, supply chains, drug and device manufacture and transport can be severely affected by extreme weather. Be it wind or wildfire events damaging buildings, practitioners migrating due to prolonged drought or roads being washed away in floods, there are a myriad of ways that healthcare can be disrupted.

For example, as we saw in Chapter 3, Hurricane Katrina in 2005 destroyed much of the New Orleans healthcare system. Even 10 years later New Orleans was still trying to recover from the effects and the number of doctors in the New Orleans area declined by almost 75% after Katrina. A US study looked at the vulnerability of healthcare to hurricanes and found that 25 of 78 on the US Atlantic and Gulf Coasts had more than half of their hospitals at risk of flooding even from relatively weak hurricanes. Only an 82cm sea level rise would increase the odds of hospital flooding by 22% and at least half of the roads within 1.6 km of hospitals were at risk of flooding from a category 2 cyclone [82].

Migration is influenced by climate but, as we have already seen, migration has healthcare costs [83]. Migrants entering a new area may have health needs that put pressure on local health services. Migrants may bring infectious diseases previously unknown in the host area or conversely suffer from the infectious diseases prevalent in the new area. Climate change can also cause health staff to migrate to less inhospitable areas. An Australian study of health professionals found that 34% of respondents indicated that climate change is already causing, or likely to cause, them to consider leaving the Northern Territory [84].

Transportation is commonly affected by weather events [85]. Transportation is vital for the functioning of healthcare. Staff and patients need to get to faculties, equipment and stores have to be transported, food has to be delivered, and engineers are needed to fix equipment, water, or electricity supplies. Climate-related events are already interrupting vital supply chains globally and these can particularly impact healthcare systems as shown following Hurricane Maria in 2017. This had a significant effect upon pharmaceutical production in Puerto Rico and produced severe shortages in the US for some months [86]. Cold-chain transport is vital for many eye drops that are used for glaucoma, such as *Bimataprost*, *Lantaprost,* and *Travaprost*, but is increasingly sensitive to supply chain disruption from climate change [87].

We have already seen those eye conditions that will be worsened by climate change and projections of how much this will cost. The Spanish study previously described, estimated that climate change–related ocular disease was already responsible for nearly 1% of Spanish GDP [40]. This money has to come from somewhere, so politicians need to decide where it comes from. All areas of healthcare will be affected by climate change so money cannot necessarily be taken from other diseases. If this money comes from social care or education funding the risks of compounding the health inequalities persist. If money is taken from infrastructure budgets how will protection and mitigation of climate damage be paid for?

References

1. World Health Organization. World Report on Vision. 2019. https://iris.who.int/bitstream/handle/10665/328717/9789241516570-eng.pdf?sequence=18.
2. World Health Organization. WHO Estimate Global Costs of Reaching Global Health Targets by 2030. WHO 2017. https://www.who.int/news/item/17-07-2017-who-estimates-cost-of-reaching-global-health-targets-by-2030.
3. Berkowitz ST, Finn AP, Parikh R et al. Ophthalmology workforce projections in the United States, 2020 to 2035. *Ophthalmology* 2024 Feb;131(2):133–139. https://doi.org/10.1016/j.ophtha.2023.09.018.
4. World Cancer Research Fund International. Skin Cancer Statistics. 2022. https://www.wcrf.org/cancer-trends/skin-cancer-statistics.
5. Cancer.net. Eyelid Cancer: Statistics. 2023. https://www.cancer.net/cancer-types/eyelid-cancer/statistics.
6. Lin HY, Cheng CY, Hsu WM, Kao WH, Chou P. Incidence of eyelid cancers in Taiwan: A 21-year review. *Ophthalmology* 2006 Nov;113(11):2101–2107. https://doi.org/10.1016/j.ophtha.2006.06.001.
7. Quigley C, Deady S, Hughes E, McElnea E, Zgaga L, Chetty S. National incidence of eyelid cancer in Ireland (2005–2015). *Eye* (Lond) 2019 Oct;33(10):1534–1539. https://doi.org/10.1038/s41433-019-0437-8.
8. Lee S-B, Saw S-M, Au Eong K-G et al. Incidence of eyelid cancers in Singapore from 1968 to 1995. *British Journal of Ophthalmology* 1999;83:595–597.
9. Safi S, Ahmadzade M, Karimi S et al. A registration trend in eyelid skin cancers and associated risk factors in Iran, 2005–2016. *BMC Cancer* 2023 Sep 30;23(1):924. https://doi.org/10.1186/s12885-023-11414-z.
10. Jung SK, Lim J, Yang SW, Jee D, Won YJ. Nationwide trends in the incidence and survival of eyelid skin cancers in Korea. *Ophthalmic Epidemiology* 2020 Dec;27(6):438–448. https://doi.org/10.1080/09286586.2020.1767152.
11. Kao SZ, Ekwueme DU, Holman DM, Rim SH, Thomas CC, Saraiya M. Economic burden of skin cancer treatment in the USA: An analysis of the Medical Expenditure Panel Survey Data, 2012–2018. *Cancer Causes & Control* 2023 Mar;34(3):205–212. https://doi.org/10.1007/s10552-022-01644-0.
12. Guy GP Jr, Machlin SR, Ekwueme DU, Yabroff KR. Prevalence and costs of skin cancer treatment in the U.S., 2002–2006 and 2007–2011. *American Journal of Preventive Medicine* 2015 Feb;48(2):183–187. https://doi.org/10.1016/j.amepre.2014.08.036.
13. Tinghög G, Carlsson P, Synnerstad I, Rosdahl I. Societal cost of skin cancer in Sweden in 2005. *Acta Dermato-Venereologica* 2008;88(5):467–473. https://doi.org/10.2340/00015555-0523.
14. Cancer.net. Eye Melanoma: An Introduction. 2023. https://www.cancer.net/cancer-types/eye-melanoma/introduction.
15. Cancer.net. Eye Melanoma: Statistics. 2023. https://www.cancer.net/cancer-types/eye-melanoma/statistics.
16. Vajdic CM, Kricker A, Giblin M et al. Incidence of ocular melanoma in Australia from 1990 to 1998. *International Journal of Cancer* 2003;105:117–122. https://doi.org/10.1002/ijc.11057.
17. Moriarty JP, Borah BJ, Foote RL, Pulido JS, Shah ND. Cost-effectiveness of proton beam therapy for intraocular melanoma. *PLoS ONE* 2015;10(5):e0127814. https://doi.org/10.1371/journal.pone.0127814.

18. WebMD. Treating Metastatic Uveal Melanoma. Marijke Durning 2023. https://www.webmd.com/melanoma-skin-cancer/treatment-metastatic-uveal-melanoma.
19. Yu GP, Hu DN, McCormick S, Finger PT. Conjunctival melanoma: Is it increasing in the United States? *American Journal of Ophthalmology* 2003 Jun;135(6):800–806. https://doi.org/10.1016/s0002-9394(02)02288-2.
20. Stapleton F, Alves M, Bunya V et al. TFOS DEWS II epidemiology report. *The Ocular Surface* 2017;15(3):334–365. https://doi.org/10.1016/j.jtos.2017.05.003.
21. Farrand KF, Fridman M, Stillman IÖ, Schaumberg DA. Prevalence of diagnosed dry eye disease in the United States among adults aged 18 years and older. *American Journal of Ophthalmology* 2017 Oct;182:90–98. https://doi.org/10.1016/j.ajo.2017.06.033.
22. McCann P, Abraham AG, Mukhopadhyay A et al. Prevalence and incidence of dry eye and Meibomian gland dysfunction in the United States: A systematic review and meta-analysis. *JAMA Ophthalmology* 2022 Dec 1;140(12):1181–1192. https://doi.org/10.1001/jamaophthalmol.2022.4394.
23. Yu D, Deng Q, Wang J et al. Air pollutants are associated with dry eye disease in urban ophthalmic outpatients: A prevalence study in China. *Journal of Translational Medicine* 2019 Feb 15;17(1):46. https://doi.org/10.1186/s12967-019-1794-6.
24. Future Market Insights. Dry Eye Syndrome Treatment Market Outlook 2023–2033. https://www.futuremarketinsights.com/reports/dry-eye-syndrome-treatment-market.
25. Chan C, Ziai S, Myageri V et al. Economic burden and loss of quality of life from dry eye disease in Canada. *BMJ Open Ophthalmology* 2021;6:e000709. https://doi.org/10.1136/bmjophth-2021-000709.
26. Yu J, Asche CV, Fairchild CJ. The economic burden of dry eye disease in the United States: A decision tree analysis. *Cornea* 2011 Apr;30(4):379–387. https://doi.org/10.1097/ICO.0b013e3181f7f363.
27. Bae S, Seung H, Oh HJ. Points to consider when developing drugs for dry eye syndrome. *Osong Public Health and Research Perspectives* 2023 Apr;14(2):70–75. https://doi.org/10.24171/j.phrp.2023.003.
28. Liu YC, Wilkins M, Kim T, Malyugin B, Mehta JS. Cataracts. *Lancet* 2017 Aug 5;390(10094):600–612. https://doi.org/10.1016/S0140-6736(17)30544-5.
29. Khairallah M, Kahloun R, Borne R et al. Vision loss expert group of the global burden of disease study; number of people blind or visually impaired by cataract worldwide and in world regions, 1990 to 2010. *Investigative Ophthalmology & Visual Science* 2015;56(11):6762–6769. https://doi.org/10.1167/iovs.15-17201.
30. Minassian DC, Mehra V. 3.8 million blinded by cataract each year: Projections from the first epidemiological study of incidence of cataract blindness in India. *British Journal of Ophthalmology* 1990 Jun;74(6):341–343. https://doi.org/10.1136/bjo.74.6.341.
31. Foster A. Vision 2020: The cataract challenge. *Community Eye Health* 2000;13(34):17–19. PMID: 17491949; PMCID: PMC1705965.
32. Rendia. Is Climate Change the Next Big Risk to Eye Health. Smitha Gopal. https://rendia.com/resources/insights/climate-change-eye-health/#:~:text=Experts%20estimate%20that%20by%202050,treatment%20of%20these%20additional%20cases.
33. Laser Eye Surgery Hub. Cataract Statistics and Data. 2023. https://www.lasereyesurgeryhub.co.uk/data/cataract-statistics/.

34. Hatch WV, Campbell EDL, Bell CM, El-Defrawy SR, Campbell RJ. Projecting the growth of cataract surgery during the next 25 years. *Arch Ophthalmology* 2012;130(11):1479–1481. https://doi.org/10.1001/archophthalmol.2012.838.
35. Marques A, Jacqueline Ramke J, John Cairns J et al. The economics of vision impairment and its leading causes: A systematic review. *eClinicalMedicine* 2022;46:101354. https://doi.org/10.1016/j. eclinm.2022.101354.
36. Wong WL, Su X, Li X et al. Global prevalence of age-related macular degeneration and disease burden projection for 2020 and 2040: A systematic review and meta-analysis. *The Lancet Global Health* 2014;2(2):e106–e116. https://doi.org/10.1016/S2214-109X(13)70145-1.
37. AMD Alliance International. The Global Economic Cost of Visual Impairment. From *Access Economics*, 2010. https://www.aa.com.tr/en/archive/amd-alliance-international-first-ever-estimates-of-global-cost-of-vision-loss-reported-today/423006.
38. Qualls LG, Hammill BG, Wang F et al. Costs of newly diagnosed neovascular age-related macular degeneration among medicare beneficiaries, 2004–2008. *Retina* 2013 Apr;33(4):854–861. https://doi.org/10.1097/IAE.0b013e31826f065e.
39. Almony A, Keyloun KR, Shah-Manek B et al. Clinical and economic burden of neovascular age-related macular degeneration by disease status: A US claims-based analysis. *Journal of Managed Care & Specialty Pharmacy* 2021 Sep;27(9):1260–1272. https://doi.org/10.18553/jmcp.2021.27.9.1260.
40. Echevarría-Lucas L, Senciales-González JM, Medialdea-Hurtado ME, Rodrigo-Comino J. Impact of climate change on eye diseases and associated economical costs. *International Journal of Environmental Research and Public Health* 2021 Jul 5;18(13):7197. https://doi.org/10.3390/ijerph18137197.
41. World Health Organization. Fact Sheets: Trachoma. 2022. https://www.who.int/news-room/fact-sheets/detail/trachoma.
42. Ferede AT, Dadi AF, Tariku A, Adane AA. Prevalence and determinants of active trachoma among preschool-aged children in Dembia District, Northwest Ethiopia. *Infectious Diseases of Poverty* 2017 Oct 9;6(1):128. https://doi.org/10.1186/s40249-017-0345-8.
43. Frick K, Keuffel E, Bowman R. Epidemiological, demographic, and economic analyses: Measurement of the value of trichiasis surgery in The Gambia. *Ophthalmic Epidemiology* 2001;8:191–201. https://doi.org/10.1076/opep.8.2.191.4163.
44. Mohammadpour M, Abrishami M, Masoumi A, Hashemi H. Trachoma: Past, present and future. *Journal of Current Ophthalmology* 2016 Sep 19;28(4):165–169. https://doi.org/10.1016/j.joco.2016.08.01.
45. Ramesh A, Bristow J, Kovats S et al. The impact of climate on the abundance of *Musca sorbens*, the vector of trachoma. *Parasites Vectors* 2016;9:48. https://doi.org/10.1186/s13071-016-1330-y.
46. World Health Organization. Fact Sheets: Onchocerciasis 2022. https://www.who.int/news-room/fact-sheets/detail/onchocerciasis.
47. Luisa F, Di Francesco G, Gianfranco Giorgio P et al. Onchocerciasis: Current knowledge and future goals. *Frontiers in Tropical Diseases* 2022:3. https://doi.org/10.3389/fitd.2022.986884.
48. Academia.edu. Climate Change and Onchocerciasis: Insights from an Analysis Public Datasets. Ronaldo Figueiró 2023. https://www.academia.edu/112476385

/Climate_change_and_onchocerciasis_insights_from_an_analysis_of_public_datasets?uc-sb-sw=39159908.

49. Kim YE, Sicuri E, Tediosi F. Financial and economic costs of the elimination and eradication of onchocerciasis (River Blindness) in Africa. *PLOS Neglected Tropical Diseases* 2015 Sep 11;9(9):e0004056. https://doi.org/10.1371/journal.pntd.0004056.
50. Brown L, Leck A, Gichangi M. The global incidence and diagnosis of fungal keratitis. *The Lancet Infectious Diseases* 2021;21(3):E49–E57. :https://doi.org/10.1016/S1473-3099(20)30448-5.
51. Radhakrishnan N, Pathak N, Subramanian KR et al. Comparative study on costs incurred for treatment of patients with bacterial and fungal keratitis – A retrospective analysis. *Indian Journal of Ophthalmology* 2022 Apr;70(4):1191–1195. https://doi.org/10.4103/ijo.IJO_2176_21.
52. Chantra S, Jittreprasert S, Chotcomwongse P, Amornpetchsathaporn A. Estimated direct and indirect health care costs of severe infectious keratitis by cultured organisms in Thailand: An 8-year retrospective study. *PLoS One* 2023 Jul 12;18(7):e0288442. https://doi.org/10.1371/journal.pone.0288442.
53. Market Research Future. Fungal Eye Infection Market. 2024. https://www.marketresearchfuture.com/reports/fungal-eye-infection-market-5498.
54. Nnadi NE, Carter DA. Climate change and the emergence of fungal pathogens. *PLOS Pathogens* 2021 Apr 29;17(4):e1009503. https://doi.org/0.1371/journal.ppat.1009503.
55. Centre for Disease Control and Prevention. Climate Changer and Fungal Disease. 2023. https://www.cdc.gov/fungal/climate.html.
56. Zhang Y, Xu X, Wei Z et al. The global epidemiology and clinical diagnosis of Acanthamoeba keratitis. *Journal of Infection and Public Health* 2023;16(6):841–852. https://doi.org/10.1016/j.jiph.2023.03.020.
57. Centre for Disease Control and Prevention. Acanthamoeba. 2021. https://www.cdc.gov/parasites/acanthamoeba/infection-sources.html#:~:text=The%20incidence%20of%20the%20disease,per%20million%20contact%20lens%20wearers.
58. UNICEF. Vitamin A Supplementation: A Decade of Progress. 2007. https://factsforlife.org/pdf/Vitamin_A_Supplementation.pdf.
59. American Academy of Ophthalmology. What is Vitamin A Deficiency? Kiersten Boyd 2023. https://www.aao.org/eye-health/diseases/vitamin-deficiency#:~:text=Vitamin%20A%20deficiency%20is%20the,year%20of%20losing%20their%20sight.
60. Wirth JP, Petry N, Tanumihardjo SA et al. Vitamin A supplementation programs and country-level evidence of Vitamin A deficiency. *Nutrients* 2017 Feb 24;9(3):190. https://doi.org/10.3390/nu9030190.
61. Neidecker-Gonzales O, Nestel P, Bouis H. Estimating the global costs of vitamin A capsule supplementation: A review of the literature. *Food and Nutrition Bulletin* 2007 Sep;28(3):307–316. https://doi.org/10.1177/156482650702800307.
62. Tang K, Eilerts H, Imohe A et al. Evaluating equity dimensions of infant and child vitamin A supplementation programmes using Demographic and Health Surveys from 49 countries. *BMJ Open* 2023;13:e062387. https://doi.org/10.1136/bmjopen-2022-062387.
63. World Health Organization. Health Topics: Physical Activity. https://www.who.int/data/gho/indicator-metadata-registry/imr-details/3416.

64. World Health Organization. Obesity and Overweight. https://www.who.int/news-room/fact-sheets/detail/obesity-and-overweight.
65. Qin L, Knol MJ, Corpeleijn E, Stolk RP. Does physical activity modify the risk of obesity for Type 2 Diabetes: A review of epidemiological data. *European Journal of Epidemiology* 2010;25(1):5–12. https://doi.org/10.1007/s10654-009-9395-y.
66. International Diabetes Federation. Facts and Figures. 2021. https://idf.org/about-diabetes/diabetes-facts-figures/.
67. Sun H, Saeedi P, Karuranga S et al. IDF diabetes Atlas: Global, regional and country-level diabetes prevalence estimates for 2021 and projections for 2045. *Diabetes Research and Clinical Practice* 2022 Jan;183:109119. https://doi.org/10.1016/j.diabres.2021.109119.
68. Diabetes.co.uk. Cost of Diabetes. https://www.diabetes.co.uk/cost-of-diabetes.html.
69. O'Connell JM, Manson SM. Understanding the economic costs of diabetes and prediabetes and what we may learn about reducing the health and economic burden of these conditions. *Diabetes Care* 2019 Sep;42(9):1609–1611. https://doi.org/10.2337/dci19-0017.
70. Yau JW, Rogers SL, Kawasaki R et al. Meta-Analysis for Eye Disease (META-EYE) Study Group. Global prevalence and major risk factors of diabetic retinopathy. *Diabetes Care* 2012 Mar;35(3):556–564. https://doi.org/10.2337/dc11-1909.
71. Thomas RL, Halim S, Gurudas S, Sivaprasad S, Owens DR. IDF diabetes Atlas: A review of studies utilising retinal photography on the global prevalence of diabetes related retinopathy between 2015 and 2018. *Diabetes Research and Clinical Practice* 2019 Nov;157:107840. https://doi.org/10.1016/j.diabres.2019.107840.
72. Shah AR, Gardner TW. Diabetic retinopathy: Research to clinical practice. *Clinical Diabetes and Endocrinology* 2017 Oct 19;3:9. https://doi.org/10.1186/s40842-017-0047-y.
73. Happich M, Reitberger U, Breitscheidel L et al. The economic burden of diabetic retinopathy in Germany in 2002. *Graefe's Archive for Clinical and Experimental Ophthalmology* 2008;246:151–159. https://doi.org/10.1007/s00417-007-0573-x.
74. Rein DB, Zhang P, Wirth KE et al. The economic burden of major adult visual disorders in the United States. *Arch Ophthalmology* 2006;124(12):1754–1760. https://doi.org/10.1001/archopht.124.12.1754.
75. Teo ZL, Tham YC, Yu M et al. Global prevalence of diabetic retinopathy and projection of burden through 2045: Systematic review and meta-analysis. *Ophthalmology* 2021 Nov;128(11):1580–1591. https://doi.org/10.1016/j.ophtha.2021.04.027.
76. Tham Y-C, Li X, Wong TY et al. Global prevalence of glaucoma and projections of glaucoma burden through 2040. *Ophthalmology* 2014;121:2081–2090.
77. Wang W, He M, Li Z, Huang W. Epidemiological variations and trends in health burden of glaucoma worldwide. *Acta Ophthalmology* 2019;97:e349–e355. https://doi.org/10.1111/aos.14044.
78. Négrel A, Thylefors B. The global impact of eye injuries. *Ophthalmic Epidemiology* 1998;5(3):143–169. https://doi.org/10.1076/opep.5.3.143.8364.
79. Swain T, McGwin G Jr. The prevalence of eye injury in the United States, estimates from a meta-analysis. *Ophthalmic Epidemiology* 2020 Jun;27(3):186–193. https://doi.org/10.1080/09286586.2019.1704794.

80. Hashemi A, Khabazkhoob M, Mehri A et al. Epidemiology of ocular trauma in the elderly: A population-based study. *Journal of Current Ophthalmology* 2023 Aug 11;35(1):79–85. https://doi.org/10.4103/joco.joco_53_23.
81. Li C, Fu Y, Liu S et al. The global incidence and disability of eye injury: An analysis from the Global Burden of Disease Study 2019. *EClinicalMedicine* 2023 Aug 9;62:102134. https://doi.org/10.1016/j.eclinm.2023.102134.
82. Tarabochia-Gast AT, Michanowicz DR, Bernstein AS. Flood risk to hospitals on the United States Atlantic and Gulf Coasts from hurricanes and sea level rise. *GeoHealth* 2022;6:e2022GH000651. https://doi.org/10.1029/2022GH000651.
83. Chen M, Caldeira K. Climate change as an incentive for future human migration. *Earth System Dynamics* 2020;11:875–883. https://doi.org/10.5194/esd-11-875-2020.
84. Pendrey C, Qulity S, Gruen R et al. Is climate change exacerbating health-care workforce shortages for underserved populations? *The Lancet Planetary Health* 2021;5(4):e183–e184. https://doi.org/10.1016/S2542-5196(21)00028-0.
85. Gössling S, Neger C, Steiger R et al. Weather, climate change, and transport: a review. *Nature Hazards* 2023;118:1341–1360. https://doi.org/10.1007/s11069-023-06054-2.change.ng
86. Lawrence J-M, Ullah Ibne Hossain N, Jaradat R, Hamilton M. Leveraging a Bayesian network approach to model and analyze supplier vulnerability to severe weather risk: A case study of the U.S. pharmaceutical supply chain following Hurricane Maria. *International Journal of Disaster Risk Reduction* 2020;49:101607. https://doi.org/10.1016/j.ijdrr.2020.101607.
87. Centre For Sustainable Cooling. Sustainable and Resilient Cold Chains: The 2050 Imperative. 2022. https://sustainablecooling.org/wp-content/uploads/2023/03/The-Local-to-Global-Summit-Report.pdf

5 Eye care's effects upon climate change

The previous chapters have shown the impact of climate change on human health and in particular the risk to our eyes and vision. But as all human activities impact upon our planet and therefore our climate, the converse also holds. Healthcare already significantly impacts our environment and eye care is a significant part of healthcare. This chapter looks at the main impacts that the management of eye conditions can have upon the climate.

5.1 The effect of healthcare on the climate

The 2019 *Lancet Countdown on Health and Climate Change* reports that the healthcare sector is responsible for almost 5% of global greenhouse gas emissions – around 2 gigatonnes of carbon dioxide equivalent (2×10^{12}kgCO_2eq) [1]. This means it has a carbon output equivalent to 514 coal-fired power plants. If the sector were a country, it would be the fifth largest polluter on Earth. If this rate were to continue, emissions from healthcare would triple by 2050 [2].

About 30% of greenhouse gas emissions come directly from the healthcare facility itself, such as electricity use, the burning of waste, and transport. The remaining 70% is more indirect and is produced mainly through the supply chains that include the production, transport and disposal of medications, food, medical devices, and hospital equipment [3].

An analysis from the UK's NHS reported that in 2019, the health service's emissions totalled 25 megatonnes of carbon dioxide equivalent (MtCO2eq). Of the 2019 footprint, 62% came from the supply chain, 24% from the direct delivery of care, 10% from staff commute and patient and visitor travel and 4% from private health and care services commissioned by the NHS [4]. Supply chain emissions were dominated by the manufacturing of goods such as pharmaceuticals and chemicals 32% (5.1 MtCO2eq) and medical equipment 19% (3 MtCO2eq). The study also showed that emissions associated with the provision of business services such as indemnity insurance was 17% (2.7 MtCO2eq) of the total [4].

DOI: 10.4324/9781003512608-6

This results in a per-capita output of 540kgCO_2eq in the NHS in England. Other healthcare systems have attempted to calculate their own outputs. A Japanese study reported 566kgCO_2eq per capita in 2015 [5], an Austrian study 799kgCO_2 per capita in 2014 [6], a Canadian study 899kgCO_2eq per capita in 2015 [7], an Australian study 1,495kgCO_2eq per capita in 2015 [8] and the US 1889kgCO_2eq per capita in 2013 [9]. Although these figures are not strictly comparable, as they are calculated in different ways, they do give some idea of relative emission amounts.

Healthcare produces an enormous amount of solid waste – especially plastics. It might be thought that waste, though responsible for a huge global pollution problem, is not linked to climate change. But it is. From production through to disposal the greenhouse gas emissions are enormous. Take, for example, single-use plastics, which are estimated from their production to emit 1.5 to 12.5 million metric tons of greenhouse gases [10]. The refinement of these plastics emits an additional 184 to 213 million metric tons of greenhouse gases each year. Landfills, where single-use plastics are sent, account for more than 15% of methane emissions [10]. It has been estimated that by the middle of the century, global emissions from plastic production could triple to account for one-fifth of the Earth's remaining carbon budget [11].

High risk healthcare waste, as it contains potential poisons (e.g. chemotherapy agents) or potentially infected matter (e.g. surgical waste), needs to be specially treated before it is disposed of. The most common methods for this are autoclaving, microwaving, or incineration, all of which are high-energy procedures. This requires an enormous amount of energy as temperatures need to be high, so greenhouse gases are emitted to provide the energy to burn and in the burning itself. Ash is produced in the burning, and this can itself cause environmental damage [12].

In the US alone it has been estimated that a hospital patient will generate about 33.8 pounds of waste each day. This means about 6 million tons of waste annually. Of the 14,000 tons of waste generated daily in US healthcare facilities, about 20% to 25% is plastic, but 91% of those plastics are not recycled [13]. Globally, Personal Protective Equipment (PPE; gowns, masks, and gloves) usage was rising even before the COVID-19 pandemic. The latter caused a global surge in usage of PPE which reached unprecedented levels. This was estimated by UNICEF to be 2.2 billion surgical masks, 1.1 billion gloves, 13 million goggles, and 8.8 million face shields during the first year of the pandemic [14]. Global production of healthcare PPE increased by approximately 300–400% during the pandemic, with the steepest increases in surgical masks and gloves [14]. In 2020 use of surgical masks and gowns in the US alone contributed the carbon dioxide equivalent of 78 coal-fired power plants running continuously [15, 16].

5.2 The effect of eye care on the climate

It is difficult to accurately separate the climate impact of eye care from the overall impact of healthcare. However, a rough estimate can be made in regard to the proportion of clinical activity from the speciality. Buchan *et al.* report that in the UK, ophthalmology is the highest volume specialty accounting for 8.1% of hospital outpatient visits in 2018–19 [17]. This means over 9 million use NHS ophthalmology outpatient services per year. Around 6% of all surgical operations in the UK are ophthalmic and the demand for ophthalmology services is forecast to increase by 40% over the next 20 years [18]. Note that 40% is an extrapolation from the current workload and ageing population and does not take into account the potential effects of climate change. We have already seen that these will put an additional strain on eye care services, which means of course greater emissions.

Work has been done on specific parts of ophthalmology, including waste and variability between different countries for the same procedures. Most of the work has been done on cataract surgery as it is one of the most common surgical operations all over the world, but evidence from other ophthalmic procedures is beginning to emerge [19].

5.3 Cataract surgery, climate change, and sustainability

There are currently few studies on the impact of eyecare on the climate. Of those there are, most work has been done on cataract surgery pathways. One such study calculated the carbon output for one cataract operation was 181.8kgCO_2eq. Building and energy use was estimated to account for 36.1% of overall emissions, travel 10.1% and procurement 53.8%, with medical equipment accounting for the most emissions at 32.6% [20]. A further study found that a single phacoemulsification operation produced 280g of plastic waste and 8g of paper waste [21].

Buchan *et al.* in their *Lancet Planetary Health* paper extrapolate these figures from the UK to the global population of 10 billion. If globally all who needed cataract surgery were to get it would generate 14.5 Mt of CO_2eq annually. This would be similar to the entire annual greenhouse gas emissions of Croatia [22].

One sobering paper from India reported that a cataract facility generated 250 grams of waste per phacoemulsification and nearly 6kg of CO_2eq per phacoemulsification. This works out as about 5% of the UK's, with comparable clinical outcomes. They found that most of the environmental emissions occurred in the sterilisation process as reusable materials were mainly used. Electricity use in the operating room and the Central Sterile Services Department (CSSD) accounted for 10% to 25% of most environmental emissions (the energy of which could be gained from renewable sources when available) [23].

A Spanish study calculated that a single cataract surgery was responsible for 86.62kg CO_2eq, of which 85% was from medical and pharmaceutical equipment [24]. A study from New Zealand found that the average emissions produced by each cataract operation was 152kg CO_2eq. This is equivalent to 62L of petrol and would take 45m^2 of forest one year to absorb. The great majority of emissions were from procurement, mostly disposable materials and the second greatest contribution was from travel (driving) [25]. A French study reported that disposable items accounted for 59.49kg (73.32%) CO_2eq for each procedure. A single procedure generated around 2.83kg of waste. The average CO_2eq produced by travel to the facility was 7.26±6.90kg (8.95%) of CO2eq. The CO_2eq produced by the sterilisation of the phacoemulsifier handpiece was 2.12kg (2.61%). The energy consumption of the building and staff transportation accounted for the remaining CO_2eq emissions, 0.76kg (0.93%) and 0.08kg (0.10%) respectively. The climate footprint of one cataract procedure was 81.13kgCO_2eq – the equivalent to an average car driving 800km [26]. A Malaysian study calculated the total waste produced from 203 phacoemulsifications and found it to be 168kg. 95kg (56.6%) was clinical waste, 63kg (37.6%) was general waste and 10kg (5.8%) was sharps. The mean waste production per case of phacoemulsification surgery for an experienced ophthalmologist was 0.814kg but 1.086kg per case for a trainee. Per case this was 0.282kgCO_2eq, so like the Indian study, this found much lower emissions per case than in Europe and the US [27].

There are other ways that waste, and therefore emissions, can be reduced in cataract surgery. For example, 32.2% of waste generated during cataract surgery comprises packaging materials for intraocular lenses (IOL) and viscoelastic. A notable contributor to this waste is the presence of information leaflets in multiple languages, which are redundant because they are rarely used and could be easily replaced by a link to a website or QR code. Whilst the IOL itself weighs less than 1g its packaging weighs a substantial 64g, including a 70-page booklet translated into 11 different languages [28].

One study calculated that across four US surgical sites, the greatest proportion of unused pharmaceuticals during phacoemulsification consisted of eyedrops, with cost estimates of as much as US$195,200 [29]. The potential monthly environmental effect from these bottles could reach carbon dioxide equivalents of 2498kg, fine particulate matter of 4.5kg and nitrogen equivalents of 0.42kg. 65.7% of eye drops were unused at time of discarding, resulting in unnecessary potential carbon emissions of 711 to 2,135kg of CO_2 equivalents per month [29]. Suggestions to reduce this include more use of multidose bottles, making sure opened drops are used to their maximum shelf-life and allowing patients to take home opened drop bottles e.g. antibiotics [30].

A potentially useful tool has been developed called *Eyefficiency*. This is a cataract surgical services auditing tool to help global units improve their surgical productivity and reduce their costs, waste generation and climate footprint [31]. One important use is to allow comparisons between different units to try to identify variations in cost and waste of each procedure.

A study used the *Eyefficiency* tool to compare nine global cataract surgical facilities. They found huge variations in cost and waste per procedure in different countries. The average per-case expenditures ranged between £31.55 (India) and £399.34 (New Zealand), with the majority of costs attributable to medical equipment and supplies. Average solid waste ranged between 0.19kg (African site) and 4.27kg (UK site) per phacoemulsification, and greenhouse gases ranged from 41kgCO_2eq (India) to 130kgCO_2eq (Hungary) per phacoemulsification. Single-use supplies (including their manufacturing and upstream logistics) made up the largest proportion of greenhouse gas emissions at four other sites (the UK, Hungary, Mexico and New Zealand), ranging from 48% to 67% per case [31].

This study indicates that emissions and waste between units in disparate countries can be measured and measurement is the first step in positive change. The variations between units are profound and need further research to bring the highest emitters more in line with the lower. It is also significant to see the variation in the costs of each procedure. Healthcare planners worldwide aim to reduce the costs of healthcare in their own areas and tools like this can be very helpful. It also indicates again to those who feel that aiming for 'net-zero emissions' is too expensive that in reality the converse holds true.

5.4 Dry eye disease, eye drop use, and climate change

Dry eye disease is an example of a chronic disease that has the potential to have a significant amount of waste and emissions. Each patient or clinical interaction or treatment might have a very minor effect, but, as it has been reported that up to 50% of people have dry eye problems, cumulatively the effects could be enormous [32]. As we have seen in the previous chapter, dry eye disease is a condition likely to be negatively affected by continuing climate change.

A study by Latham *et al.* looked at the dry eye pathway and how it contributes to climate change [33]. Management of dry eye inevitably involves large amounts of medications – usually eye drops in non-recyclable bottles – which require supply chains including cold-chain. Healthcare appointments are recurrent, involving travel and energy consumption of the facility. They discuss how each step can be analysed to reduce waste, emissions and cost [33].

Lubricating eye drops generate significant plastic waste, with recycling rates for these plastic containers less than 8.7% [34]. To reduce this, single drop plastic bottles could be avoided and multidose bottles can be used instead. As the latter can last up to 28 days they are saving at least 27 throw-away plastic bottles. Development of devices that allowed prolonged contact of the lubricants (real or artificial) on the ocular surface reduce the number of drops needed in a day and therefore reduce plastic waste. Promoting preservative-free eye drops avoids the use of harmful preservatives like benzalkonium chloride. Eye drops result in an annual CO_2 impact of 7kg per person, mainly due to plastic production and waste processing [35]. More work needs to be done regarding the recycling of packaging for medications and devices. This is true for all conditions and pharmaceutical and device companies need to address this issue.

Reducing travel to hospitals can be addressed by using more local providers (e.g. community optometrists or more teleophthalmology). The latter is becoming more acceptable to practitioners and patients (accelerated by COVID-19) and will be an important strategy for cost control and waste reduction in the future. For all travel elements – from patient transport to delivery of pharmaceuticals – the use of electric vehicles can have a significant impact and needs to be encouraged.

Somner *et al.* looked at the use of diagnostic eye drops in an ophthalmology clinic in the UK. To avoid cross-infection these are single-use, meaning they are individually packaged and thrown away after each use. Their study did show that cross-contamination could occur using single-use drops if used in more than one patient. They calculated that, if 10 million patients are seen per annum by NHS ophthalmology services, and if diagnostic drops were to be reused, this would lead to a cost saving of £2.75 million and a reduction in waste of 6.85 tonnes of paper waste and 12.69 tonnes of plastic waste. Conversely, around 32 000 patients would be put at risk of cross-contamination with *Staphylococcus aureus*, meaning around £87 pounds saved for each potential cross-contamination event. Waste and emission reductions do come at a cost, and this must be recognised. Financially this cost saving of multi-use drops is significant, but it needs to be balanced with the acceptance from patient and practitioner of the small risk of iatrogenic infection [36].

As Latham *et al.* suggest, government, national and regional health systems, healthcare governing bodies, universities, independent sector companies, healthcare professionals and patients have a duty to minimise environmental waste [33]. There are numerous opportunities to improve environmental sustainability of the dry eye disease care pathway that can be explored without jeopardising the safety of patients or the financial position of organisations.

5.5 The retina and climate change

As with so many eye diseases we saw in the previous chapter the retina is vulnerable to the changes occurring in our climate – in particular, age-related macular degeneration (ARMD), which is the leading cause of irreversible blindness in those over 65. So-called neovascular or wet ARMD can now be treated with intravitreal injections of anti-VEGFs and these can stabilise visual loss. This has meant an exponential rise in these injections, so much so that many eye departments have had to redesign their whole departments to deal with them. Inevitably this means more pharmaceutical costs and waste, disposables such as gowns, masks, syringes, and instruments and travel to healthcare facilities, as follow up occurs every 1–2 months.

These numbers are only going to increase as the populations age and the effects of climate change increase. Additionally, dry ARMD intravitreal treatments are being trialled and if successful will mean even more attendances and treatments.

A study from Ireland looked at the emissions associated with intravitreal injections. They found that the emission from a single intravitreal injection was 13.68kgCO_2eq. Patient travel accounted for the majority of emissions at 77%, with procurement accounting for 19% and building energy usage for 4% of total emissions. A rationalisation of the injection pack contents would reduce carbon emissions by an estimated 0.56kg per injection. Note that this calculation did not include the actual anti-VEGF – its manufacture, transport, or disposal – so the CO_2eq figure above will be an underestimate [37].

A similar study from New Zealand looked at the environmental cost of a single intravitreal injection. When the costs of materials, travel, building energy, and waste disposal were combined the emissions were calculated as 14.1kgCO_2eq [38].

The anti-VEGFs mentioned above are of various types. Some are longer lasting than others and because of decreased travel and hospital use the emissions associated with them will be lower. Some intravitreal implants can work for months and so for the same reasons will have lower overall emissions associated with them. A Dutch study found that moving from an 8 week injection regime to a 16 week regime reduced the CO_2eq by around 40% [39].

A US study looked at the solid waste from intravitreal injections. All the waste from 337 consecutive intravitreal injections was collected. This totalled 65.6kg of waste. 83% was from cardboard boxes, foam coolers, cold packs and bubble wrap. The rest of the waste was from nitrile gloves, tissues, wipes, plastic or paper packaging and biohazard waste (used syringes and needles). Cold packs, foam coolers, cardboard/paper and nitrile gloves were the greatest contributors to carbon emissions and landfill [40].

Clinical investigations produce their own emissions. A study by Reynolds *et al.* looked at the difference between the emissions of two retinal blood vessel investigations – fluorescein angiography (IVFA) and ocular coherence topography angiography (OCTA). They found that using OCTA rather than IVFA saved 80.51 CO_2eq per patient. OCTA needed fewer visits and did not need any pharmaceuticals. Additional greenhouse emissions associated with IVFA compared to OCTA were calculated as: building costs 40.17kgCO2eq, staff travel 33.03kgCO_2eq, patient travel 4.27kgCO_2eq, pharmaceuticals 1.01kgCO_2eq, medical instruments 1.41kgCO_2eq and waste 1.24kgCO_2eq [41].

Retinal surgery often requires expansile gases to tamponade the retina. One of these is called SF6 and is the most potent greenhouse gas regulated by the Kyoto protocol with a global warming potential of 22,800 relative to CO_2.

A UK study calculated that the use of these gases in just their unit was the equivalent of 2.7 metric tons of CO_2eq emissions per year. This gas is used in retinal surgery all over the world with little recognition of its greenhouse potential [42].

5.6 Glaucoma and climate change

The main surgical intervention for glaucoma currently is trabeculectomy. A study compared the waste from trabeculectomy surgery performed in India with that from the same operation in the US [43]. The average waste from India was 0.5 ± 0.2kg significantly lower than that from the US facility, which was 1.4 ± 0.4kg per trabeculectomy. They also compared waste when glaucoma surgical devices were used or when phacoemulsification was added to the trabeculectomy with averages of 0.4 ± 0.2kg for India and 0.7 ± 0.2kg for the US. The study noted the greater use of reusable equipment in the Indian setting and of using eyedrops bottles in multiple patients. Note that the infection rates were no different in the two centres [43].

Patients with glaucoma represent a large number of ophthalmic outpatient consultations – with more than one million glaucoma-related visits just in the UK [44]. Emissions arise from the large amount of travel and the maintenance of the health facility. Diagnostic tests produce waste, including single-use plastics. A study from Boston found that single-use tonometer prisms produced around 100.8kg of plastic waste per year. Single-use gonioscopy lenses produced 8.8kg of plastic waste per year at the same facility. They calculated from these figures that 109.6kg of plastic waste could be avoided each year by only using non-disposable tonometer prisms and gonioscopy lenses [45].

As we saw in the dry eye section, plastic eyedrop bottles represent a significant amount of waste. Glaucoma is initially treated using daily eyedrops and these are used for life. A single patient may use 2–3 different bottles a day and these are not recyclable as they contain pharmaceuticals. Some patients require preservative-free drops, and the use of a single drop container is far more wasteful than multidrop bottles [46]. Each bottle with a 10ml capacity weighs 6.5g, but when the same treatment is in a preservative-free single disposable vial it generates 120g a month. Practitioners can reduce this waste by avoiding overprescribing and using multidose bottles and patients can help by avoiding wasting drops or discarding before they are finished. For those who have trouble instilling drops and consequently often waste significant amounts, eye drop aids should be used.

Increasingly, a specific laser is being used as a first-line glaucoma treatment (Selective Laser Trabeculoplasty). As this can negate the use of eyedrops in many patients it has the potential to significantly reduce waste and emissions in the longer term [47].

5.7 Optometry and climate change

Eyecare does not only happen in hospitals and community optometrists provided over 13 million sight tests in the UK in 2019/2020, as well as other diagnostic tests [48]. The Centre for Sustainable Healthcare and the NHS looked at five optometry practices over one year. This amounted to 25,745 sight tests and the calculated emissions for this totalled 135,573kgCO_2eq. 69% of this was from travel, 24% energy use in the building, 11% procurement and 5% from waste. The average emission associated with a sight test was 5.27kgCO_2eq with a range from 4.02kgCO_2eq to 9.28kgCO_2eq. This range would suggest that optometry – like ophthalmology – already has the ability to reduce emissions simply by analysing the systems in lower emitting practices [49].

The figures above were calculated for the test alone and did not include the emissions involved in making spectacles or using contact lens. When the study added these in they calculated:

kgCO_2eq Sight Test Only	5.27
kgCO_2eq Sight Test and One pair of Spectacles	8.64
kgCO_2eq Sight test and monthly contact lenses	40.24
kgCO_2eq Sight test and 6 months of daily disposable contact lenses	49.24

Contact lenses have a significant environmental impact. For example, in 2019 there were more than 45 million contact lens wearers in the US and 35% to 46% wore daily disposable contact lenses. Each contact lens weighs 30 micrograms and the use of contact lens products comprised 0.5% of the total environmental waste. Amongst 400 contact lens wearers surveyed, 19% discarded their contact lenses into the toilet or sink. On an annual basis, this results in an estimated 2.5 billion contact lenses weighing approximately 44,000 pounds entering wastewater treatment plants in the US [50].

Though less directly polluting than contact lenses, spectacles have their own issues. The plastic frames are made from heavily laminated acetates which are derived from non-renewable oil. Their manufacture is highly polluting with some containing lead and mercury. Most cannot at present be recycled so generally end in landfill where they breakdown very slowly [51].

One possible solution to the above is refractive laser eye surgery. This can mean no further contact lenses are needed and, in the under 50s, no spectacles. Over a lifetime this has the ability to significantly reduce waste from discarded contact lenses and spectacles [52].

5.8 Practitioner awareness of climate change

We have seen that eye care, ophthalmology, optometry, pharmaceutical, and medical device companies all contribute to the climate footprint. Increasing research has been done in this area – the major emitters, why there is such a disparity between different systems and countries and what can be done. However, only a few areas of eye care have currently been addressed and much more research is needed – on how it affects the climate and the balance between the risks to the climate versus the treatment. The most important thing all of us can do is to constantly try to avoid waste – be it energy or plastic packaging. As we might think of the cost effectiveness or the risk benefit balance of a treatment we should increasingly be thinking of its environmental impacts.

Climate change can affect almost everything and almost everything can affect climate change. All practitioners need to know and understand the dangers and how they can help mitigate these dangers. A paper by Chang and Thiel in the *Journal of Cataract and Refractive Surgery* in 2020 indicates that awareness of waste and sustainability is already widespread in healthcare practitioners. They conducted an online survey of more than 1300 cataract surgeons and nurses and 93% believed that operating room waste is excessive and should be reduced. 78% believed that more supplies should be reusable. 90% were concerned about global warming and 87% wanted medical societies to advocate for reducing the surgical carbon footprint [53].

Chandra *et al.* conducted a similar study amongst New Zealand ophthalmologists. They reported as part of their results that 19% of respondents expressed the opinion that climate change was not due to human activity and did not require mitigation. Younger ophthalmologists tended to have greater agreement with the need for broad-based political action on climate mitigation than those aged over 50 years. Most practices reported that they had room to improve on reducing waste, travel and carbon footprints [54].

Finally, Cameron *et al.* from the UK Health Foundation commissioned a survey of 1,858 adults in 2021. They were asked questions about climate change, health, and the responsibility of the NHS. Most respondents were concerned about the health impacts of climate change. 6% opposed the NHS's ambition to attain Net Zero by 2045, though only 26% had realised that the NHS made a significant contribution to climate change. When questioned more deeply, 91% would return unused medications and 85% would accept reusable equipment. Fewer would consider the environmental impact when deciding on their own treatment (64%). This falls even further when they were asked about a direct impact on individual patient care. 30% of people would not be willing to consider the environmental impact of their treatment, with opposition increasing to 34% for people older than 65. Likewise, 30% would not opt for a remote GP consultation (to reduce travel, rather than seeing a doctor in person), increasing to 40% for those aged 55 and older [55].

Overall, the public and practitioner awareness and understanding of climate change is high. Most seem to agree that climate change can impact health and therefore needs to be addressed. These numbers are encouraging but words now need to be put into action.

References

1. Watts N, Amann M, Arnell N et al. The 2019 report of The Lancet Countdown on health and climate change: Ensuring that the health of a child born today is not defined by a changing climate. *Lancet* 2019 Nov 16;394(10211):1836–1878. https://doi.org/10.1016/S0140-6736(19)32596-6.
2. The Health Policy Partnership. The Healthcare Sector's Environmental Impact. 2022. https://www.healthpolicypartnership.com/the-nexus-between-climate-change-and-healthcare/#:~:text=The%20healthcare%20sector%27s%20environmental%20impact,fifth%20largest%20polluter%20on%20Earth.
3. Healthcare without Harm. Healthcares Climate Footprint. 2019. https://noharm-global.org/sites/default/files/documents-files/5961/HealthCaresClimateFootprint_092319.pdf.
4. Tennison I, Roschnik S, Ashby B et al. Health care's response to climate change: A carbon footprint assessment of the NHS in England. *The Lancet Planetary Health* 2021;5(2):e84–e92. https://doi.org/10.1016/S2542-5196(20)30271-0.ed.
5. Nansai K, Fry J, Malik A, Takayanagi W, Kondo N. Carbon footprint of Japanese health care services from 2011 to 2015. *Resources, Conservation and Recycling* 2020;152:104525. https://doi.org/10.1016/j.resconrec.2019.104525.
6. Weisz U, Pichler P-P, Jaccard IS et al. Carbon emission trends and sustainability options in Austrian health care. *Resources, Conservation and Recycling* 2020;160:104862. https://doi.org/10.1016/j.resconrec.2020.104862.
7. Eckelman MJ, Sherman JD, MacNeill AJ. Life cycle environmental emissions and health damages from the Canadian healthcare system: An economic-environmental-epidemiological analysis. *PLOS Medicine* 2018;15(7):e1002623. https://doi.org/10.1371/journal.pmed.1002623.
8. Malik A, Lenzen M, McAllister S, McGain A. The carbon footprint of Australian health care. *The Lancet Planetary Health* 2018;2(1):e27–e35. https://doi.org/10.1016/S2542-5196(17)30180-8.
9. Eckelman MJ, Sherman JD. Estimated global disease burden from US health care sector greenhouse gas emissions. *American Journal of Public Health* 2018 Apr;108(S2):S120–S122. https://doi.org/10.2105/AJPH.2017.303846.
10. University of Colorado. The Impact of Plastic on Climate Change. 2023. https://www.colorado.edu/ecenter/2023/12/15/impact-plastic-climate-change#:~:text=The%20refinement%20of%20plastics%20emits,landfill%20size%20and%20these%20emissions.
11. The Guardian. Plastic Production Emissions Could Triple to One-Fifth of the Earth's Carbon Budget. https://www.theguardian.com/us-news/2024/apr/18/plastic-production-emission-climate-crisis.
12. Zikhathile T, Atagana H, Bwapwa J, Sawtell D. A review of the impact that healthcare risk waste treatment technologies have on the environment. *International Journal of Environmental Research and Public Health* 2022 Sep 22;19(19):11967. https://doi.org/10.3390/ijerph191911967.
13. AMA Journal of Ethics. Should US Health Care Lead Global Change in Plastic Waste Disposal? 2022. https://journalofethics.ama-assn.org/article/how-should-us-health-care-lead-global-change-plastic-waste-disposal/2022-10.
14. UNICEF. COVID-19 Impact Assessment and Outlook on Personal Protective Equipment. https://www.unicef.org/supply/stories/covid-19-impact-assessment-and-outlook-personal-protective-equipment.

15. Bromley-Dulfano R, Chan J, Jain N, Marvel J. Switching from disposable to reusable PPE. *BMJ* 2024 Mar 18;384:e075778. https://doi.org/10.1136/bmj-2023-075778.
16. Uddin MA, Afroj S, Hasan T et al. Environmental impacts of personal protective clothing used to combat COVID-19. *Advanced Sustainable Systems* 2022 Jan;6(1):2100176. https://doi.org/10.1002/adsu.202100176.
17. International Agency for the Prevention of Blindness. Sustainability in Eye Care: More Evidence Needed to Reduce Environmental Impact. Dr John Buchan 2022. https://www.iapb.org/blog/sustainability-in-eye-care/.
18. The Royal College of Ophthalmologists. New RCOphth Workforce Census Illustrates the Severe Shortage of Eye Doctors in the UK. 2019. https://www.rcophth.ac.uk/news-views/new-rcophth-workforce-census-illustrates-the-severe-shortage-of-eye-doctors-in-the-uk/.
19. Sherry B, Lee S, Ramos Cadena MLA et al. How ophthalmologists can decarbonize eye care: A review of existing sustainability strategies and steps ophthalmologists can take. *Ophthalmology* 2023 Jul;130(7):702–714. https://doi.org/10.1016/j.ophtha.2023.02.028.
20. Morris DS, Wright T, Somner JE, Connor A. The carbon footprint of cataract surgery. *Eye* (Lond) 2013 Apr;27(4):495–501. https://doi.org/10.1038/eye.2013.9.
21. Somner J, Scott K, Morris D, Gaskell A, Shepherd I. Ophthalmology carbon footprint: Something to be considered? *Journal of Cataract & Refractive Surgery* 2009;35:202–203.
22. Buchan JC, Thiel CL, Steyn A et al. Addressing the environmental sustainability of eye health-care delivery: A scoping review. *Lancet Planet Health* 2022 Jun;6(6):e524–e534. https://doi.org/10.1016/S2542-5196(22)00074-2.
23. Thiel CL, Schehlein E, Ravilla T et al. Cataract surgery and environmental sustainability: Waste and lifecycle assessment of phacoemulsification at a private healthcare facility. *Journal of Cataract & Refractive Surgery* 2017 Nov;43(11):1391–1398. https://doi.org/10.1016/j.jcrs.2017.08.017.
24. Pascual-Prieto J, Nieto-Gómez C, Rodríguez-Devesa I. The carbon footprint of cataract surgery in Spain. *Archivos de la Sociedad Española de Oftalmología* (Engl Ed). 2023 May;98(5):249–253. https://doi.org/10.1016/j.oftale.2023.01.005.
25. Latta M, Shaw C, Gale J. The carbon footprint of cataract surgery in Wellington. *The New Zealand Medical Journal* 2021 Sep 3;134(1541):13–21. PMID: 34531593.
26. Ferrero A, Thouvenin R, Hoogewoud F et al. The carbon footprint of cataract surgery in a French University Hospital. *Journal Français d'Ophtalmologie* 2022 Jan;45(1):57–64. https://doi.org/10.1016/j.jfo.2021.08.004.
27. Khor HG, Cho I, Lee KRCK, Chieng LL. Waste production from phacoemulsification surgery. *Journal of Cataract & Refractive Surgery* 2020 Feb;46(2):215–221. https://doi.org/10.1097/j.jcrs.0000000000000009.
28. The American Academy of Ophthalmology. Eye Wiki. The Environmental Sustainability of Cataract Surgery. 2023. https://eyewiki.org/The_Environmental_Sustainability_of_Cataract_Surgery#.
29. Tauber J, Chinwuba I, Kleyn D et al. Quantification of the cost and potential environmental effects of unused pharmaceutical products in cataract surgery. *JAMA Ophthalmology* 2019 Oct 1;137(10):1156–1163. https://doi.org/10.1001/jamaophthalmol.2019.2901.

30. Healio. Ocular Surgery News. Needless Drug Waste in Ophthalmology: What Can Be done? https://www.healio.com/news/ophthalmology/20240208/needless-drug-waste-in-ophthalmology-what-can-be-done#:~:text=Pharmaceutical%20waste%20in%20ophthalmic%20surgical,lightly%20used%20before%20being%20discarded.
31. Goel H, Wemyss TA, Harris T et al. Improving productivity, costs and environmental impact in International Eye Health Services: Using the 'Eyefficiency' cataract surgical services auditing tool to assess the value of cataract surgical services. *BMJ Open Ophthalmology* 2021 May 20;6(1):e000642. https://doi.org/10.1136/bmjophth-2020-000642.
32. Stapleton F, Alves M, Bunya V et al. TFOS DEWS II epidemiology report. *The Ocular Surface* 2017;15(3):334–365. https://doi.org/10.1016/j.jtos.2017.05.003.
33. Latham SG, Williams RL, Grover LM, Rauz S. Achieving net-zero in the dry eye disease care pathway. *Eye (Lond)* 2024 Apr;38(5):829–840. https://doi.org/10.1038/s41433-023-02814-3.
34. Govindasamy G, Lim C, Riau AK, Tong L. Limiting plastic waste in dry eye practice for environmental sustainability. *The Ocular Surface* 2022 Jul;25:87–88. https://doi.org/10.1016/j.jtos.2022.05.005.
35. The American Academy of Ophthalmology. Eye Wiki. The Environmental Sustainability of Dry Eye Disease. 2023. https://eyewiki.aao.org/The_Environmental_Sustainability_of_Dry_Eye_Disease#cite_note-:24-25.
36. Somner JE, Cavanagh DJ, Wong KK et al. The precautionary principle: What is the risk of reusing disposable drops in routine ophthalmology consultations and what are the costs of reducing this risk to zero? *Eye (Lond)* 2010 Feb;24(2):361–363. https://doi.org/10.1038/eye.2009.129.
37. Power B, Brady R, Connell P. Analyzing the carbon footprint of an intravitreal injection. *Journal of Ophthalmic and Vision Research* 2021 Jul 29;16(3):367–376. https://doi.org/10.18502/jovr.v16i3.9433.
38. Chandra P, Welch S, Oliver GF, Gale J. The carbon footprint of intravitreal injections. *Clinical & Experimental Ophthalmology* 2022 Apr;50(3):347–349. https://doi.org/10.1111/ceo.14055.
39. ISPOR Europe 2023. The Environmental Impact of Aflibercept for the Treatment of Neovascular Age Related Macular Degeneration and Diabetic Macular Oedema in the Netherlands (Poster). https://www.ispor.org/docs/default-source/euro2023/ispor2023environmentalimpactafli2910131553-pdf.pdf?sfvrsn=eed1090e_0.
40. Cameron TW 3rd, Vo LV, Emerson LK et al. Medical waste due to intravitreal injection procedures in a retina clinic. *Journal of VitreoRetinal Diseases* 2021 Feb 10;5(3):193–198. https://doi.org/10.1177/2474126420984657.
41. ARVO Meeting Abstract 2019. The Carbon Footprint of Fluorescein Angiography Compared to OCT Angiography. R Reynolds, D Morris, U Chakravarthy. https://iovs.arvojournals.org/article.aspx?articleid=2743008.
42. Chadwick O, Cox A. Response to Tetsumoto et al. regarding the use of fluorinated gases in retinal detachment surgery. The environmental impact of fluorinated gases. *Eye (Lond)* 2021 Oct;35(10):2891. https://doi.org/10.1038/s41433-020-01197-z.
43. Namburar S, Pillai M, Varghese G, Thiel C, Robin AL. Waste generated during glaucoma surgery: A comparison of two global facilities. *American Journal of Ophthalmology Case Reports* 2018 Oct 10;12:87–90. https://doi.org/10.1016/j.ajoc.2018.10.002.

44. Fight for Sight. Back a Breakthrough in Glaucoma. https://www.fightforsight.org.uk/our-research/glaucoma/#:~:text=It%20is%20estimated%20that%2C%20by,to%20UK%20hospitals%20each%20year.
45. Park EA, LaMattina KC. Economic and environmental impact of single-use plastics at a large ophthalmology outpatient service. *Journal of Glaucoma* 2020 Dec;29(12):1179–1183. https://doi.org/10.1097/IJG.0000000000001655.
46. Govindasamy G, Lim C, Riau AK, Tong L. Limiting plastic waste in dry eye practice for environmental sustainability. *The Ocular Surface* 2022 Jul;25:87–88. https://doi.org/10.1016/j.jtos.2022.05.005.
47. Gazzard G, Konstantakopoulou E, Garway-Heath D et al. LiGHT Trial Study Group. Selective laser trabeculoplasty versus eye drops for first-line treatment of ocular hypertension and glaucoma (LiGHT): A multicentre randomised controlled trial. *Lancet* 2019 Apr 13;393(10180):1505–1516. https://doi.org/10.1016/S0140-6736(18)32213-X.
48. Statista: State of Health. Number of NHS Sight Tests in England from 2000/01 to 2019/20. https://www.statista.com/statistics/430616/eyesight-tests-through-the-national-health-service-in-england-uk/.
49. NHS and The Centre for Sustainable Health Care. The Annual Carbon Footprint of NHS Sight Test at Five Optometry Practices. https://networks.sustainablehealthcare.org.uk/sites/default/files/resources/The%20Annual%20Carbon%20Footprint%20of%20NHS%20Sight%20Tests%20at%20Five%20Optometry%20Practices_1.pdf.
50. Contact Lens Spectrum. The Environmental Impact of Contact Lens Waste. 2019. https://clspectrum.com/issues/2019/august/the-environmental-impact-of-contact-lens-waste/#reference-11.
51. Hansaj R, Govender B, Joosah M et al. Spectacle frames: Disposal practices, biodegradability and biocompatibility – A pilot study. *African Journal Vision and Eye Health* 2021;80(1):a621. https://doi.org/10.4102/aveh.v80i1.621.
52. My-iclinic. The Financial, Environmental and Lifestyle Benefits of Laser Eye Surgery. https://www.my-iclinic.co.uk/articles/the-financial-environmental-and-lifestyle-benefits-of-laser-eye-surgery.
53. Chang DF, Thiel CL, Ophthalmic Instrument Cleaning and Sterilization Task Force. Survey of cataract surgeons' and nurses' attitudes toward operating room waste. *Journal of Cataract & Refractive Surgery* 2020 Jul;46(7):933–940. https://doi.org/10.1097/j.jcrs.0000000000000267.
54. Chandra P, Gale J, Murray N. New Zealand ophthalmologists' opinions and behaviours on climate, carbon and sustainability. *Clinical & Experimental Ophthalmology* 2020 May;48(4):427–433. https://doi.org/10.1111/ceo.13727.
55. The Health Foundation. Going Green: What Do the Public Think about the NHS and Climate Change? https://www.health.org.uk/publications/long-reads/going-green-what-do-the-public-think-about-the-nhs-and-climate-change.

6 The burden of blindness

Becoming blind is a tragedy for an individual but also resonates through families, communities, and health and social care systems. To have an economically inactive individual in poorer communities has real significance and real impacts. There is a cost to that individual and to the support systems they have around them.

We have seen that climate change is going to increase eye disease – both numbers and severity. This inevitably will mean more people will become blind and therefore unable to participate fully in their communities. We are heading for a situation where increasing visual impairment exists in a more hostile world, where in certain regions food and clean water are becoming more and more scarce. As ever this burden will initially be in the economically developing countries around the equator – countries which already have the least resistance to climate change and the least ability to support those who are impaired by it.

6.1 What is blindness?

This is perhaps harder to define than might be assumed. Someone could read all the way down a vision chart but have such a degree of peripheral visual field loss that they would struggle to navigate a street, e.g. glaucoma. Similarly, someone else may not be able to read a newspaper but have good peripheral vision, e.g. macular disease.

The use of different terms can be confusing and can vary between countries. The standard international definition of blindness is Snellen acuity of less than 3/60. However, India uses visual acuity levels less than 6/60 to define blindness, as do some African and Chinese systems. The US also uses 6/60 as the cut-off for the definition [1].

It is important to try to reach standard definitions so numbers can be compared between countries and over time. The generally accepted way to do this is via the WHO *International Classification of Diseases* (ICD) – the current version is ICD-11 [2]. The full ICD-11 categories for visual impairment can be found in the appropriate section on the ICD-11 site [3]. The categories go

DOI: 10.4324/9781003512608-7

from mild visual impairment, through moderate to severe visual impairment to different degrees of blindness. There has been some criticism of these categories [4]. For example, they are defined as when refractive error is corrected but, as we will see in the next section, many people do not have their refractive error corrected. If your visual impairment isn't corrected (e.g. because of a lack of facility for eye testing or an inability to afford glasses) then you have the same level of disability as someone with the same vision level that can't be corrected.

6.2 Current numbers and projections

The WHO provides an extensive report on the prevalence of visual impairment and blindness globally [5].

They estimate that at least 2.2 billion people have a near or distance vision impairment. Of these, half – 1 billion – could have had this impairment prevented (e.g. trachoma or onchocerciasis) or are in areas where treatment is not accessible (e.g. cataract or refractive error).

The main conditions causing loss of distance vision are: cataracts (94 million), refractive errors (88.4 million), age related macular degeneration (8 million), glaucoma (7.7 million) and diabetic retinopathy (3.9 million). The main condition causing near visual impairment is presbyopia (826 million), a condition which is fully correctable with extremely cheap reading glasses.

The WHO report states that the prevalence of distance vision impairment in low- and middle-income regions is four times higher than in high-income regions. Rates of unaddressed near vision impairment are estimated to be greater than 80% in western, eastern, and central sub-Saharan Africa, while comparative rates in high-income regions of North America, Australasia, Western Europe and Asia-Pacific are reported to be lower than 10%.

The WHO also report that globally only 36% of people with a distance vision impairment due to refractive error have received access to an appropriate pair of spectacles and only 17% of people with vision impairment or blindness due to cataract have received access to quality surgery.

A major review of global blindness numbers and trends estimated that in 2020 43.3 million people were blind. Globally, between 1990 and 2020, amongst adults aged 50 years or older, age-standardised prevalence of blindness decreased by 28.5% and prevalence of mild vision impairment decreased slightly by –0.3%, whereas prevalence of moderate and severe vision impairment increased slightly by 2.5% [6].

They estimated that the largest number of people with blindness were in South Asia, followed by East Asia and South-East Asia. The crude prevalence of blindness ranged from 1.94 cases per 1000 in high-income North America to 8.75 cases per 1000 in South-East Asia. The age-standardised prevalence of blindness ranged from 1.24 cases per 1000 people in high-income North America to 11.1 per 1000 in western sub-Saharan Africa [6].

They extrapolated their findings to predict that by 2050, 61 million people would be blind, 474 million have moderate and severe vision impairment, 360 million will have mild vision impairment and 866 million (629 to 1150) will have uncorrected presbyopia. Therefore, even though age prevalence slightly declined, overall numbers increased and are set to increase further. This may be a function of an enlarging population (especially increases where there is more uncorrected visual impairment) and/or at least partly caused by climate change [6].

When looking at years lived with disability (YLD) data for 2019, blindness and vision loss resulted in 22.6 million global YLDs in 2019, a 20.3% (18.9–21.9) increase since 2010. When ranked against all causes of disease by YLDs in the *Global Burden of Diseases, Injuries and Risk Factors Study* 2019, blindness and low vision ranked eighth in the 50–69 years age group for both sexes and fourth in the aged 70 years and older age group [6, 7]

The International Agency for the Prevention of Blindness (IAFB) predict that the numbers of people with visual loss will increase to 1,758 million by 2050. By 2045, 161 million will have diabetic retinopathy 45 million of whom it will be sight-threatening. In 2015 it was estimated that 23% of the world's population were short-sighted but this is projected to increase to 50% by 2050 [8].

6.3 The economic costs

The WHO *World Report on Vision* describes the economics of visual impairment [9]. A study from nine countries estimated that the annual cost of moderate to severe vision impairment ranged from US$0.1 billion in Honduras to as high as US$16.5 billion in the US. Annual global costs of productivity losses associated with vision impairment from uncorrected refractive error were estimated to be around US$270 billion. Uncorrected myopia in the regions of East Asia, South Asia and South-East Asia were reported to be more than twice that of other regions and equivalent to more than 1% of gross domestic product.

A study from Gordon *et al.* calculated the worldwide cost of vision loss as an estimated US$3 trillion. Direct health costs were US$2.3 trillion whilst the remaining indirect costs were due to lost productivity, premature death, and the value of informal caregivers. They broke this down further and estimated the global financial cost of uncorrected refractive error was US$1.6 trillion and of age-related macular degeneration (AMD) was US$0.3 trillion [10]. Glaucoma, which is a leading cause of blindness in African Americans, costs the U.S. economy US$2.86 billion every year in direct costs and productivity losses [11].

The IAFB reported that the worldwide economic burden attributed to vision impairment and blindness is US$411 billion annually which is equivalent to roughly 0.3% of the global GDP. People with sight loss experience a 30% relative reduction in employment and have a higher mortality risk. Educational achievement is reduced with implications for future earning potential. Conversely, providing corrective spectacles increases work productivity by 22% and every US$1 spent on eyecare can yield US$36 of economic return [12].

The International Labour Organization reported that around the world, over 13 million people live with vision impairment linked to their work, with an estimated 3.5 million eye injuries occurring in the workplace every year. This amounts to 1 per cent of all non-fatal occupational injuries [13].

Though the estimates from different studies vary it is beyond doubt that visual impairment and blindness cost. They cost the individual in educational and occupational opportunities. They cost that person's family as they need to make up for these lost costs as well as supporting the individual. They cost society both from the money needed to support the impaired person (optical, social support, daily living support) and the loss of revenue from that economically inactive person.

6.4 The personal costs

Behind each of the statistics listed above is a person struggling to survive and becoming increasingly dependent upon others. Visual impairment can have profound effects upon an individual's social interactions, mental health, occupational opportunities, and financial situation.

The UK organisation Fight For Sight produced a report called *Time to Focus* [14]. They found that workforce barriers and informal care for those with sight loss had an economic cost of £25.2 billion a year. This cost they predicted would rise to £33.5 billion by 2050. 84% of this was made up of informal care by family and friends, productivity losses, and quality of life impacts. 37% of those of working age with severe sight loss were not working due to their eye conditions and overall 25% felt they were struggling financially.

The report also found that quality of life of people with severe sight loss was lower than that of people with depression, arthritis, and even advanced breast cancer. 70% of those who were visually impaired felt that some area of their life was limited by their eye condition and that it had a negative impact on their personal relationships. The mental health impact of sight loss was highest in those with a low annual household income, adding to anxiety and depression. 37% found the emotional challenges of living with their conditions hardest and 30% felt their eye condition limited their independence and freedom to make choices. Overall, the report found that the perceived levels of loneliness were significantly higher than in the general population.

One US study found that visual impairment cost US$4,000 per person per year whilst a different study put this at nearly US$12,000 per year [15]. This is of course in a country where eyecare is universally available. In many countries it is not – often in regions with the greatest demand. Ironically, the direct costs of visual impairment in these countries will be lower simply because the person does not have access to eyecare services and so cannot spend that money. This inevitably means greater costs fall upon family and neighbours who must support that person.

The WHO report illustrates this – direct costs, including costs involved in accessing eye care, transport to appointments and related pharmaceutical interventions, are primary barriers to accessing care, particularly in low- and middle-income countries. The main explanation given for this is that approximately 50% of people in low- and middle-income countries live more than one hour from a city (compared with 10% in high-income countries), so transport and its costs become a major barrier to accessing to eye care services [9].

Rein *et al.* found that in Americans aged 65 years or older, 16% of those who are visually impaired and 40% of those who are blind reside in nursing homes, compared with only 4.3% of those in the general population. This they estimated meant that 424,801 more visually impaired and blind Americans were in nursing homes than would be expected if these same individuals had normal vision. This means that visual impairment and blindness accounted for US$11 billion in nursing home costs in 2004. They also calculated the direct costs of guide dogs to be US$62 million [16].

Finally, a study from Boagey *et al.* looked at the psychological and mental health aspects of visual loss. They found high levels of depression, stress and anxiety, and even suicidal ideation directly linked to the person's vision. All these seem to be underpinned by a sense of loss and the process of trying to create a new identity as a visually impaired person [17].

This chapter has presented the global numbers of those whose daily lives are affected by visual impairment. The numbers are huge even in developed economies and we can see that they represent another area where climate change can take its toll. Through the diseases we have already looked at we can see this toll may already be beginning and the sooner we address the climate crises the better for all of us.

References

1. Dandona L, Dandona R. What is the global burden of visual impairment? *BMC Medicine* 2006 Mar 16;4:6. https://doi.org/10.1186/1741-7015-4-6.
2. World Health Organization. International Statistical Classification of Diseases and Related Health Problems 2022. https://www.who.int/standards/classifications/classification-of-diseases.

3. ICD-11 for Mortality and Morbidity Statistics. 9D90 Vision Impairment Including Blindness. https://icd.who.int/browse/2024-01/mms/en#1103667651.
4. Dandona L, Dandona R. Revision of visual impairment definitions in the international statistical classification of diseases. *BMC Medicine* 2006 Mar 16;4:7. https://doi.org/10.1186/1741-7015-4-7.
5. World Health Organization. Blindness and Visual Impairment. 2023. https://www.who.int/news-room/fact-sheets/detail/blindness-and-visual-impairment.
6. GBD 2019 Blindness and Vision Impairment Collaborators; Vision Loss Expert Group of the Global Burden of Disease Study. Trends in prevalence of blindness and distance and near vision impairment over 30 years: An analysis for the Global Burden of Disease Study. *The Lancet Global Health* 2021 Feb;9(2):e130–e143. https://doi.org/10.1016/S2214-109X(20)30425-3.
7. GBD 2019 Diseases and Injuries Collaborators. Global burden of 369 diseases and injuries in 204 countries and territories, 1990–2019: A systematic analysis for the Global Burden of Disease Study 2019. *Lancet* 2020 Oct 17;396(10258):1204–1222. https://doi.org/10.1016/S0140-6736(20)30925-9.
8. International Agency for the Prevention of Blindness. Projected Changes in Vision Loss 2020–2050. https://www.iapb.org/learn/vision-atlas/magnitude-and-projections/projected-change/.
9. World Health Organization. World Report on Vision. 2019. https://www.who.int/publications/i/item/9789241516570.
10. ARVO Annual Meeting. The Global Cost of Visual Loss (Abstract). Gordon et al. https://iovs.arvojournals.org/article.aspx?articleid=2361599.
11. Bright Focus Foundation: National Glaucoma Research. Glaucoma: Facts and Figures. https://www.brightfocus.org/glaucoma/article/glaucoma-facts-figures.
12. International Agency for the Prevention of Blindness. Eye Health and Economic Development. https://www.iapb.org/learn/vision-atlas/economics/eye-health-and-economic-development/.
13. International Labour Organization. Eye Health and the World of Work. 2023. https://www.ilo.org/global/topics/safety-and-health-at-work/resources-library/publications/WCMS_892937/lang--en/index.htm.
14. Fight for Sight. Time to Focus Report: The Future of Sight Loss. https://www.fightforsight.org.uk/our-research/timetofocus/.
15. Frick KD, Gower EW, Kempen JH, Wolff JL. Economic impact of visual impairment and blindness in the United States. *Arch Ophthalmology* 2007 Apr;125(4):544–550. https://doi.org/10.1001/archopht.125.4.544.
16. Rein DB, Zhang P, Wirth KE et al. The economic burden of major adult visual disorders in the United States. *Arch Ophthalmology* 2006 Dec;124(12):1754–1760. https://doi.org/10.1001/archopht.124.12.1754.
17. Boagey H, Jolly JK, Ferrey AE. Psychological impact of vision loss. *Journal of Mental Health and Clinical Psychology* 2022;6(3):25–31. https://doi.org/10.29245/2578-2959/2021/3.1256.

7 Rays of hope

7.1 Organisations, governments, and industry

There is growing awareness that the Earth's climate is changing and that this will harm us. Governments and organisations are starting to take action to ameliorate this damage. The Paris Agreement in 2015 aimed to reduce emissions by 45% by 2030 and to reach Net Zero by 2050 [1].

As all aspects of healthcare are going to be profoundly impacted by climate change, a number of organisations have set their own targets. The UK NHS has set two targets for *direct* emissions from the NHS to reach Net Zero by 2040, with an ambition to reach an 80% reduction between 2028 and 2032. For emissions they can *indirectly* influence (e.g. outside suppliers) to reach Net Zero by 2045, with an ambition to reach an 80% reduction between 2036 and 2039 [2]. Plans that aim to decarbonise NHS care cover a number of aspects, including construction of Net Zero hospitals, adding zero-emission vehicles to the NHS's fleet, installing LED light bulbs and optimising the location of care (e.g. moving care closer to home and reducing hospital visits and travel emissions).

At COP26 in 2021, fifty countries committed to creating climate resilient, low carbon, sustainable health systems, including 14 countries that set a target date of reaching net zero emissions by 2050. In addition to this, 54 institutions from 21 countries that represented more than 14,000 hospitals committed to achieving Net Zero emissions [3].

The International Agency for the Prevention of Blindness, recognising the threat that climate change brings to vision, has produced 10 key areas of action. It is a useful guide for any healthcare organisation, charity, government, or individuals. The 10 elements are [4]:

1. Lead – acknowledge the problem, engage, create, and monitor a strategy.
2. Advocate – health practitioners to use their powerful voices.
3. Procure sustainably.
4. Reduce the use of fossil fuels.
5. Conserve water.
6. Reduce and safely dispose of waste.

DOI: 10.4324/9781003512608-8

7. Reduce and green travel.
8. Follow the four principles of sustainable clinical practice.
9. Embed environmental sustainability in education.
10. Focus your research into environmentally sustainable eyecare.

On the commercial side more companies are realising the importance of planetary health and making firm commitments. Examples include the pharmaceutical companies AstraZeneca, GSK, Merck KGaA, Novo Nordisk, Roche, Samsung Biologics and Sanofi who in 2022 announced joint action to achieve near-term emissions reduction targets and accelerate the delivery of Net Zero health systems [5]. The imaging company Canon Medical Systems have pledged to be carbon Net Zero by 2040 [6]. Providence Healthcare aims to be carbon negative by 2030 [7].

EyeSustain is a global coalition of eye societies, organisations and ophthalmologists who are collaborating to make ophthalmic care and surgery more sustainable [8]. The International Agency for the Prevention of Blindness has set up a Climate Action Working Group [9]. The goal for this is to provide leadership and advocacy on climate action in eyecare. The Funder Commitment on Climate Change was launched in 2019 to support charitable funders in tackling the causes and impacts of climate change [10]. The EyeSustain website has a large number of resources, links, and information.

We have seen in the earlier chapters the vast burden that eye disease puts upon the individual and the society they live in. We have also seen the deleterious effects that climate change is having on these diseases. The rest of this chapter will look at some examples where this disease burden is being addressed and sometimes reversed. It will also look at some of the tools being developed to make eye care more sustainable. These examples can provide a blueprint to use in adaptions to diseases and changes that systems can make to become more robust. There are other equally good examples from all over the world and hopefully these will increase as we face the threats climate change brings.

7.2 Trachoma

The WHO have targeted the elimination of trachoma by 2030. They adopted the 'SAFE strategy' (see below) in 1993 and launched the WHO Alliance for the Elimination of Trachoma in 1996 [11].

In 1998 the World Health Assembly resolution called for trachoma elimination by 2020 using the SAFE strategy. This is an acronym for Surgical treatment, Antibiotic treatment for acute infection, Face washing and Environmental changes to improve sanitation [12]. The WHO Data report in 2021 indicated that 69,266 people with trachomatous trichiasis were provided with corrective surgery in that year and 64.6 million people in endemic communities were treated with antibiotics to eliminate trachoma. In 2019, 92,622 people with trachomatous trichiasis were provided with corrective surgery, and 95.2 million people were treated with antibiotics [13].

As of 5 October 2022, 15 countries – Cambodia, China, Gambia, Islamic Republic of Iran, Lao People's Democratic Republic, Ghana, Malawi, Mexico, Morocco, Myanmar, Nepal, Oman, Saudi Arabia, and Vanuatu – had been validated by WHO as having eliminated trachoma as a public health problem [11]. Benin and Mali were added to this list in 2023 [14].

7.3 Onchocerciasis

Since 1974 the Onchocerciasis Control Programme (OCP) in West Africa has attempted to bring the disease under control. Initially this was through vector control of the responsible fly but more recently large-scale distribution of Ivermectin has supplemented this [15]. The WHO report that the programme cured 40 million people of infection thus preventing blindness in 600,000 people and that 18 million children were born free from the threat of the disease and blindness.

In Africa in 2023, Niger became the first African country to eliminate the transmission of river blindness, and Senegal is on track to become the second country to achieve this [16]. South America, Colombia, and Ecuador were reported as having eliminated onchocerciasis by the WHO in 2013 and 2014 respectively. Mexico and Guatemala were also able to stop transmission in 2011 [15, 17].

7.4 Vitamin A deficiency

High-dose vitamin A supplementation has been shown to reduce all-cause mortality by 12 to 24 per cent. Globally, vitamin A supplementation amongst children under 5 years of age with one dose has increased from 50% to 66% [18]. Measles is a risk factor for blindness from vitamin A deficiency but in 2022, 74% of children received both doses of the measles vaccine, and about 83% of the world's children received one dose of measles vaccine by their first birthday [19].

In 1993, the Nepalese government initiated a National Vitamin A Programme (NVAP) with support from UNICEF. Amongst other positive parameters, the program reached millions of children each year and its impact was shown to have prevented blindness in around 2,000 children annually. It was also found to reduce the mortality rate for children under the age of 5 in Nepal by around 50% between 1995 and 2000 [20].

Overall, vitamin A deficiency is reducing in Asia. It also seems to be reducing in Latin America and the Caribbean. However, it does seem to be static in sub-Saharan Africa [21]. Vitamin A deficiency and blindness had decreased worldwide but much more needs to be done. Treatment is very cheap and very effective but reaching the populations at most risk is the strategy for the future.

7.5 Diabetic retinopathy

A seminal study – the Diabetes Control and Complications Trial (DCCT) – followed Type 1 diabetic patients with mild or no retinopathy for a mean of 6.5 years. They found that intensive blood sugar control reduced the adjusted mean risk for the development of diabetic retinopathy by 76% [22]. Use of statins has been shown to reduce the risk of diabetic retinopathy and the need for retinal laser [23]. Regular physical activity has also been shown to reduce the risk of retinopathy [24]. Reducing body weight by only 5–10% can significantly reduce the risk of diabetic complications [25].

Although diabetics and diabetic complications are increasing worldwide, studies do show that lifestyle changes (rather than expensive laser or injections) can have a profound effect on retinopathy. These lifestyle changes are open to anyone in the world with diabetes and require education and communication.

7.6 Technology

The *Eyefficiency* tool is mentioned in Chapter 5 and describes itself as a 'web-based toolkit which estimates the triple bottom line (social, environmental and ecological) of cataract and intravitreal surgeries using data collected in theatres via a mobile app' [26]. To know if we are making sustainable progress we need to be able to measure with a standard tool – within procedures and between organisations and countries. Mostly used for cataract surgery it is able to calculate the CO_2eq of the procedure. Some of the outcomes have highlighted the huge variation in emissions from one unit or country compared to another. Once this variation is identified the cause of this variation can be investigated and then, hopefully, be used to reduce the emissions from the higher unit [27].

Telemedicine has great potential to reduce travel and therefore emissions. One study found that its use could decrease patient travel by over 25% [28]. A US telemedicine study found a total emission reduction of 618,772kgCO_2e which represents 0.306 metric tons of CO_2e per patient [29]. A systemic review looking across all medical procedures found an average reduction in emissions of 0.70–372kgCO_2e per telemedicine review consultation [30].

The use of Artificial Intelligence in healthcare also has the potential to reduce emissions in the healthcare sector. It can be used to rationalise treatment, meaning fewer patient visits, it can reduce waste by more accurate ordering or prescribing, it can be used to compare and understand different emissions amounts from different units [31, 32]. One study estimated that AI had the potential to reduce emissions by 80% in an example they used [33].

However, AI produces its owns greenhouse gas emissions via its servers and this needs to be taken into account when assessing its potential value [34].

7.7 Education

Perhaps this intervention has the greatest potential for change. Every person on our planet will have to increasingly think of how much they take from the Earth's resources. Healthcare is no different and both practitioners and patients need to consider the impact of the healthcare they provide or the healthcare they consume.

As a paper by Tennision *et al.* looking at healthcare's response to climate change states:

> Health promotion and disease prevention programmes such as public health campaigns and social prescribing can reduce the overall demand for healthcare, while the selection of a less carbon-intensive and resource-intensive care practices where clinically appropriate can reduce both emissions and costs. The activity-based emissions results could be used to set priorities in disease prevention, as well as development and adoption of lower carbon best practices. [35]

The Global Climate and Healthcare Alliance in 2022 released an open letter calling on 'all educational stakeholders to ensure health professionals are prepared to identify, prevent and respond to the health impacts of climate change and environmental degradation. Preparing health professionals for a future impacted by climate change is the responsibility of all education stakeholders, including the deans, academics, managers and other teaching staff of health professional educational institutes as well as the associated accrediting, examination and licensing bodies' [36].

The American Academy of Ophthalmology is pushing for a more sustainable future. They are increasingly educating members to reduce waste in their everyday clinical practice, as well as advocating for climate-friendly regulations and legislation [37]. They now have a chapter in their book *Basic Principles of Ophthalmic Surgery* on reducing waste in surgical practice ensuring quality care while minimising waste and emissions [38]. In the UK Education for Sustainable Healthcare has produced a curriculum for the UK. This is aimed at all health professionals in the UK including medical students [39]. The Medical Colleges of Australia and New Zealand – including that of ophthalmology – have produced a report with resources to help healthcare professionals contribute to a more sustainable future, including advocacy [40].

It is obvious that healthcare professionals are in a position to lead the debate on climate change and health. They generally have the trust of the public, have a science background, and can directly see the harm environmental degradation can do to patients. It is refreshing that practitioners and their organisations are beginning to take this lead.

7.8 Optometry

There are a number of initiatives within optometry that are aimed at a more sustainable future. For example, Mita sustainable eyewear uses recycled plastic water bottles to make spectacle frames [41]. In the UK, Peep Eyewear recycles discarded spectacles and plants a tree for each pair purchased [42].

IbisVision is a company that provides remote platforms for optometry services. This includes an at home refraction system that substantially reduces patient travelling and therefore emissions [43].

At present, contact lens plastic can be difficult to recycle. However, biodegradable plastics are being developed, as well as systems to degrade some of the older plastics. Johnson & Johnson Vision reports that 100% of its electricity comes from renewable sources. The company recycles 89% of raw materials on location at its manufacturing sites. Further, as part of their Acuvue contact lens recycling programme, more than 8.8 million contact lenses and blister packs have been recycled in the UK alone. Alcon prevented 4,900 tons of non-recycled hazardous waste by recycling and prevention, and the company recycled 79% of all operational waste. The company reported using 10% renewable electricity and 14% recycled water in its operations, and 98% of solvents used in the company's operations are recycled offsite [44].

7.9 Finally

That our world is changing is a consistent theme of this book. There is little doubt that this change will continue and likely accelerate. That these changes will affect our health is also without doubt and our eyes will literally and figuratively be on the front line of the effects of these changes. Few, if any, eye conditions will not be worsened by our increasingly challenging climate and although this will eventually threaten all of us, initially it will threaten those with least resilience.

The recognition of the above is growing and with it the need to do something – from the national and international actions leading to Net Zero to individuals who look at their lives and try to make them more sustainable. We must all work together to deal with this threat of our own creation. We are all responsible for the state of the planet and are equally responsible for correcting it. Whilst this book finishes on an optimistic note by highlighting positive things that are being done, we all must realise that we still have a long way to go. At least though, we are now beginning to have some guides as to how to best tread this path.

References

1. United Nations: Climate Action. The Paris Agreement. https://www.un.org/en/climatechange/paris-agreement.
2. NHS England. Delivering a 'Net Zero' National Health Service. https://www.england.nhs.uk/greenernhs/wp-content/uploads/sites/51/2022/07/B1728-delivering-a-net-zero-nhs-july-2022.pdf.
3. Wise J. COP26: Fifty countries commit to climate resilient and low carbon health systems. *BMJ* 2021 Nov 9;375:n2734. https://doi.org/10.1136/bmj.n2734.
4. The International Agency for the Prevention of Blindness. 10 Key Areas of Action. https://www.iapb.org/learn/knowledge-hub/elevate/climate-action/10-key-areas-of-action/.
5. Astra Zeneca. Seven Pharma CEOs Announce New Joint Action to Accelerate Net Zero Healthcare. 2022. https://www.astrazeneca.com/media-centre/articles/2022/seven-pharma-ceos-announce-new-joint-action-to-accelerate-net-zero-healthcare.html.
6. Canon Medical. A Green Guide for UK Diagnostic Imaging. https://uk.medical.canon/wp-content/uploads/sites/8/2021/09/Green_Guide.pdf.
7. Providence. Toward Carbon Negative. https://www.providence.org/about/advocacy-and-social-responsibility/environmental-stewardship/carbon-negative-goal.
8. EyeSustain. https://eyesustain.org/about.
9. The International Agency for the Prevention of Blindness. Climate Action. https://www.iapb.org/connect/work-groups/climate-action-old/.
10. Funder Commitment on Climate Change. https://fundercommitmentclimatechange.org.
11. World Health Organization. Fact Sheets: Trachoma. 2022. https://www.who.int/news-room/fact-sheets/detail/trachoma.
12. Lavett DK, Lansingh VC, Carter MJ et al. Will the SAFE strategy be sufficient to eliminate trachoma by 2020? Puzzlements and possible solutions. *Scientific World Journal* 2013 May 19;2013:648106. https://doi.org/10.1155/2013/648106.
13. *The* International Agency for the Prevention of Blindness. WHO Reports Continued Progress toward Trachoma Elimination. 2023. https://www.iapb.org/news/who-reports-continued-progress-towards-trachoma-elimination/.
14. World Health Organization. WHO Congratulates Benin and Mali for Eliminating Trachoma as Public Health Problem. https://www.who.int/news/item/16-05-2023-who-congratulates-benin-and-mali-for-eliminating-trachoma-as-a-public-health-problem.
15. World Health Organization. Control of Neglected Diseases: Onchocerciasis. https://www.who.int/teams/control-of-neglected-tropical-diseases/onchocerciasis/prevention-control-and-elimination.
16. Bill and Melinda Gates Foundation. Ending Neglect, Ending Disease: River Blindness in West Africa. https://www.gatesfoundation.org/ideas/articles/how-to-eliminate-river-blindness-onchocerciasis-africa.
17. Lakwo T, Oguttu D, Ukety T, Post R, Bakajika D. Onchocerciasis elimination: Progress and challenges. *Research and Reports in Tropical Medicine* 2020 Oct 7;11:81–95. https://doi.org/10.2147/RRTM.S224364.

18. The International Agency for the Prevention of Blindness. Vitamin A Deficiency. https://www.iapb.org/learn/knowledge-hub/eye-conditions/vitamin-a-deficiency/.
19. World Health Organization. Measles. https://www.who.int/news-room/fact-sheets/detail/measles#:~:text=All%20children%20or%20adults%20with,prevent%20eye%20damage%20and%20blindness.
20. World Forgotten Children Foundation. How Successful Implementation of Vitamin A Led to Reduced Child Mortality in Nepal. 2022. https://www.worldforgottenchildren.org/blog/how-successful-implementation-of-vitamin-a-led-to-reduced-child-mortality-in-nepal/153#:~:text=The%20program%20reached%20millions%20of,2000%20(Gottlieb%2C%20n.d.).
21. Hamer DH, Keusch GT. Vitamin A deficiency: Slow progress towards elimination. *The Lancet Global Health* 2015 Sep;3(9):e502–3. https://doi.org/10.1016/S2214-109X(15)00096-0.
22. Diabetes Control and Complications Trial Research Group; Nathan DM, Genuth S, Lachin J et al. The effect of intensive treatment of diabetes on the development and progression of long-term complications in insulin-dependent diabetes mellitus. *The New England Journal of Medicine* 1993 Sep 30;329(14):977–986. https://doi.org/10.1056/NEJM199309303291401.
23. Simó R, Hernández C. What else can we do to prevent diabetic retinopathy? *Diabetologia* 2023 Sep;66(9):1614–1621. https://doi.org/10.1007/s00125-023-05940-5.
24. Yan X, Han X, Wu C et al. Effect of physical activity on reducing the risk of diabetic retinopathy progression: 10-year prospective findings from the 45 and Up Study. *PLoS One* 2021 Jan 14;16(1):e0239214. https://doi.org/10.1371/journal.pone.0239214.
25. O'Brien R, Johnson E, Haneuse S et al. Microvascular outcomes in patients with diabetes after bariatric surgery versus usual care: A matched Cohort Study. *Annals of Internal Medicine* 2018 Sep 4;169(5):300–310. https://doi.org/10.7326/M17-2383.
26. Eyefficiency. https://eyefficiency.org.
27. Goel H, Wemyss TA, Harris T et al. Improving productivity, costs and environmental impact in International Eye Health Services: Using the 'Eyefficiency' cataract surgical services auditing tool to assess the value of cataract surgical services. *BMJ Open Ophthalmology* 2021 May 20;6(1):e000642. https://doi.org/10.1136/bmjophth-2020-000642.
28. Wootton R, Bahaadinbeigy K, Hailey D. Estimating travel reduction associated with the use of telemedicine by patients and healthcare professionals: Proposal for quantitative synthesis in a systematic review. *BMC Health Services Research* 2011 Aug 8;11:185. https://doi.org/10.1186/1472-6963-11-185.
29. Whetten J, Montoya J, Yonas H. ACCESS to better health and clear skies: Telemedicine and greenhouse gas reduction. *Telemedicine and e-Health* 2019 Oct;25(10):960–965. https://doi.org/10.1089/tmj.2018.0172.
30. Purohit A, Smith J, Hibble A. Does telemedicine reduce the carbon footprint of healthcare? A systematic review. *Future Healthcare Journal* 2021 Mar;8(1):e85–e91. https://doi.org/10.7861/fhj.2020-0080.
31. Chakraborty C, Pal S, Bhattacharya M, Islam MA. AI-enabled ChatGPT's carbon footprint and its use in the healthcare sector: A coin has two sides. *International*

Journal of Surgery 2024 Feb 1;110(2):1306–1307. https://doi.org/10.1097/JS9.0000000000000905.

32. Bloomfield P, Clutton-Brock P, Pencheon E et al. Artificial Intelligence in the NHS: Climate and emissions. *The Journal of Climate Change and Health* 2021;4:100056. https://doi.org/10.1016/j.joclim.2021.100056.
33. Wolf RM, Abramoff MD, Channa R et al. Potential reduction in healthcare carbon footprint by autonomous artificial intelligence. *NPJ Digital Medicine* 2022 May 12;5(1):62. https://doi.org/10.1038/s41746-022-00605-w.
34. Stanford Universities Human Centred Artificial Intelligence. AI's Carbon Footprint Problem. https://hai.stanford.edu/news/ais-carbon-footprint-problem.
35. Tennison I, Roschnik S, Ashby B et al. Health care's response to climate change: A carbon footprint assessment of the NHS in England. *Lancet Planet Health* 2021 Feb;5(2):e84–e92. https://doi.org/10.1016/S2542-5196(20)30271-0.
36. The Global Climate and Health Alliance. A Call to Strengthen Climate Change Education for All Health Care Professionals. https://climateandhealthalliance.org/initiatives/who-cs-wg-call-to-strengthen-climate-change-education/.
37. American Academy of Ophthalmology. Sustainability in Ophthalmology. https://www.aao.org/sustainability.
38. American Academy of Ophthalmology. Eyenet Magazine. Making Ophthalmology More Sustainable. Reena Mukamal. https://www.aao.org/eyenet/article/making-ophthalmology-more-sustainable.
39. Medical Schools Council. Education for Sustainable Healthcare. https://www.medschools.ac.uk/media/2949/education-for-sustainable-healthcare_a-curriculum-for-the-uk_20220506.pdf.
40. University of Otago. Actions on Climate Change by Medical Colleges and Dental Associations of Australia and New Zealand. 2022. https://ranzco.edu/wp-content/uploads/2022/07/How-Colleges-Face-a-Sustainable-Future52.pdf.
41. Mita Sustainable Eyewear. https://mita-eyewear.com/pages/about.
42. Peep. Sustainable Preloved Vintage Glasses. https://www.peepeyewear.co.uk.
43. IbisVision. https://www.ibis.vision/about-us.
44. Contact Lens Spectrum. The Environmental Impact of Contact Lens Waste. https://clspectrum.com/issues/2019/august/the-environmental-impact-of-contact-lens-waste/#reference-11.

Index

Printed in the United States
by Baker & Taylor Publisher Services